Introduction to Engineering Economy For Business and Personal Applications

First Edition

Salvador Rojas-Murillo
Hofstra University

cognella
SAN DIEGO

Bassim Hamadeh, CEO and Publisher
John Remington, Managing Executive Editor
Gem Rabanera, Senior Project Editor
Susana Christie, Senior Developmental Editor
Alia Bales, Production Manager
Emely Villavicencio, Senior Graphic Designer
JoHannah McDonald, Licensing Associate
Natalie Piccotti, Director of Marketing
Kassie Graves, Senior Vice President, Editorial
Jamie Giganti, Director of Academic Publishing

Printed in the United States of America.

cognella® | ACADEMIC PUBLISHING
3970 Sorrento Valley Blvd., Ste. 500, San Diego, CA 92121

Table of Contents

Chapter 1. Introduction to Engineering Economics

Paula's College Story

IMG 1.1

Five years ago, Paula was a successful high school student who decided to pursue a career in engineering but was undecided about which engineering discipline to choose, where she should apply, and how she could pay for college. Paula had some money saved, and her family also offered to help her with some of the costs. However, she knew that getting a degree was going to be expensive as she would need money for living expenses, transportation, and school supplies. Paula's plan also included receiving a scholarship, having a part-time job, and applying for different grants. She thought that, if needed, she could also apply for student loans.

After graduating from high school Paula decided to pursue a four-year engineering degree at a state university in a neighboring state. Going to the nearby state university was cheaper than going to her home state, even when she was charged out-of-state tuition. Unfortunately for Paula, her family was unable to provide the financial support they had planned, so Paula ended up acquiring a $10,000 loan for her first year of college. In the following three years Paula acquired three more loans: $13,000 for her second year, $16,000 for her third year, and $19,000 for her fourth and final year. Paula had to acquire this debt because she was unable to work as much as she anticipated; she had a difficult time balancing work, study, and extra-curricular activities.

Although her final debt reached $58,000, Paula considered her debt "manageable" and hoped that she was going to be able to pay off her debt in a "few" years after graduation. Her university website described that engineers from her engineering discipline were hired within a few months after graduation with an average yearly salary of $70,000, so her hope was that she would fall in that average. Unfortunately, at the time Paula acquired her student loans, she did not pay much attention to the fact that her student loan had an interest rate of 7.5% per year, and that she had to start making payments one year after her graduation. She also overlooked the fact that her student loans had high late-payment fees. Luckily for Paula, her university had a great co-op program, and she received a job offer before graduating with a salary above $70,000. Just as Paula, many college students acquire student loans to finance their college education. Unfortunately, like Paula, many students acquire debt without proper financial knowledge and even graduate college without a proper financial education.

Reading Activity: Responding to Paula's College Story

- From the reading about Paula, identify the relevant information she used in her decision making to acquire these loans.
- Should Paula have borrowed the student loans? How can she know this was the correct decision financially speaking?
- What other information do you think is needed to determine if Paula made the correct decision?
- Find two similarities and two differences between borrowing to open a business and for acquiring student loans.

Learning Outcomes

After covering this chapter, the student should be able to:

1. Learn the definition of engineering economy.
2. Understand why engineering economy is a decision-making tool.
3. Learn where engineers may apply engineering economy.
4. Learn when to perform an engineering economic study and the steps of an engineering economy project.
5. Learn how calculate estimates and how to classify costs.
6. Understand the concept definition of financial illiteracy and financial impotence.
7. Recognize the relevance of engineering economy in their personal and professional life.
8. Establish the relationship between engineering economy and ethics.

Introduction: An Engineering Economy Is a Decision-Making Tool

Engineering economy is a decision-making tool based on economic analysis. As described in the reading, we all make multiple economic decisions in our daily life, and many of these may seem irrelevant because we may not feel the immediate financial impact. For instance, buying a coffee for $3.50, acquiring extended coverage on a new $1,000 smartphone for $10 dollars a month, or signing up for a new streaming service for "just a few dollars" a month could have long-term impacts on our finances. Businesses face similar situations when buying equipment, when signing a new lease, or when expanding their operations. Therefore, individuals and businesses should consider the financial implications of their decisions.

Engineering Economy Definition

Engineering economy can be defined as: a series of mathematical techniques that allow us to formulate, calculate, and determine the financial outcome of a financial problem. Moreover, given different alternatives, it allows us to determine the optimal alternative from a financial perspective.

Engineering economy considers cash flows and the interest rates that make these flows equivalent. We will go over the financial equivalence in more detail in the following pages. However, it is important to note that engineering economy does not consider market economics. In other words, we are concentrated on problems related to the time equivalence of money and not on the economy. For example, market economics would define the price of an electric vehicle based on market information such as supply and demand, and government policy, among other factors, whereas and engineering economy consider the monthly payment for buying such a vehicle or the rate of return of using this type of vehicle for a delivery service.

When to Perform an Engineering Economic Study

Based on the decision-making example, when would it be appropriate to do an economic analysis? Should we buy coffee every day from a local coffee parlor or get extended coverage on a smartphone? As in many life situations, engineering economic problems present various levels of difficulty.

In general, we may divide problems into three difficulty levels: low-difficulty, intermediate-difficulty, and high-difficulty (complex) problems.

Low-difficulty problems can be done with simple calculations or no calculations at all. For instance, deciding to buy a cup of coffee from business A versus business B solely on the price does not need complex calculations. Similarly, one could select a credit card simply by comparing interest rates. A $1 coffee is cheaper (and financially speaking better) than a $3 coffee. Similarly, a credit card charging a 25% interest rate per year (APR) is more expensive and a worse finance option than a credit card charging 10% APR. However, in the real situation some additional factors could also be considered such as: the financial capacity of the person and the benefits provided by the credit card such as miles or discounts.

On the other hand, low-difficulty problems can easily transform into an intermediate-difficulty problem. The main difference between low- and intermediate-difficulty problems is that intermediate problems consider more information and require calculations to be solved. For instance, you could perform the analysis for buying a cup of coffee every day versus buying a coffee machine. For this problem one should consider the cost of buying a coffee over a period of time versus the machine cost, duration, capacity, and the cost of coffee. Similarly, for the credit card selection problem, one credit card may have a higher interest rate but may allow the user to obtain financial benefits that the low-interest card may not such as miles, discounts, or preferable concert seats. Furthermore, a person with a credit card balance (debt) may transfer his/her balance to a high interest-rate credit card offering a promotional interest rate. This is the simple part of the problem as with no calculations he/she selects the card charging a lower interest rate. However, if the person is not able to pay the balance before the expiration date of the promotional rate, may pay higher interest rates on the new card than they would have with the original card.

The difference between intermediate- and high-difficulty (complex) problems is also based on considering more information and requiring additional analyses. High-difficulty problems are often multifactorial and go beyond financial analysis, considering factors such as tax incentives, resources, political stability, and environmental impact, among many others. For instance, in 2019 Amazon was in the process of selecting the location of their new headquarters. Such a decision required information and analyses about taxes, tariffs, permits, tax incentives, environmental considerations, salaries, housing, and logistics, among many others.

The concepts described in this text consider the solution of intermediate-difficulty problems. Some typical examples are the calculation of loan payments, the rate of return on an investment, comparison, and selection between alternatives, and plans for machine replacement or life retirement.

Exercise
Consider that your university is developing a new recreational center and you and your friends would like to open a Juice Bar at the new recreational center. Describe what type of information you would need to know to select if this business would generate a high profit.

Why Should Engineers Learn Engineering Economy?

Engineering courses can be split into two main categories: core courses, which are the foundation of engineering such as calculus and physics, and discipline-specific courses, such as heat transfer for mechanical engineers, structures for civil engineers, and operations research for industrial engineers. Since engineering economy is neither a core engineering course nor an engineering specialty course for most engineering degrees, students often wonder why they need to take this course.

For other students, this is a required course (that they would prefer to avoid), others think it will be an easy course that could help improve their GPA, and others see this course as a good opportunity to acquire some financial knowledge that they may apply for planning their future.

On the other hand, from the personal perspective, acquiring debt is far easier than paying for it. The notion that just having a high-paying job is enough to cover our financial needs is unfortunately far from true. Therefore, having healthy personal finances and generating personal wealth are the result of having good control over your personal finances.

Engineering students would benefit from acquiring engineering economy knowledge for the following reasons:

- They be prepared to solve the engineering economy section of the Principles and Practice of Engineering (PE) exam.
- They be able to prepare a financial analysis for professional (business) applications.
- They be able to perform a personal financial analysis and benefit from increasing personal wealth and minimizing financial costs.

Relevance of Engineering Economy in Technical and Personal Applications

As previously described, engineering courses (core and discipline specific) allow engineers to develop and solve engineering problems/projects such as automating an assembly line or building a bridge. However, engineering projects could fail if they are not supported by a proper financial analysis. A dramatic statistic shows that about "82% of businesses that fail do so because of cash flow problems" [1]. Without knowing the story behind these failing businesses, we could reasonably assume that for some of these projects the lack an of appropriate financial plan could have been a reason for their failure.

To clarify this point, let us consider the following examples. The first example is a technical example. In June of 1869, Thomas Edison patented a vote-recorder machine called the "*electrographic vote-recorder*". His invention's objective was to find an efficient way to record legislators' votes by flipping a switch instead of tallying up the votes by hand. Unfortunately, he failed to realize that for legislators having a manual vote provided them the opportunity to talk to their counterparts and convince them to change their vote. Consequently, having a quick and efficient way to vote would avoid such political practice and the machine was not attractive to the legislators, resulting in a financial loss. Therefore, we may infer that developing a technological advancement or technical solution to an existing problem does not always guarantee financial success. In addition, knowing the market needs is only one part of a business plan, which needs to be complemented by other components such as a marketing plan, a technical plan, and a financial plan.

The next examples come from personal finance. One could consider that financial success is only dependent on income. However, earning more money does not necessarily equate to financial stability. For example, an article by Forbes [2] described that one of the reasons why about 80% of retired NFL players go bankrupt as early as after three years out of the league is the lack of financial planning, even when they received an average salary of almost $2 million a year. Furthermore, this situation does not only apply to professional athletes. The Federal Reserve generates a report called the "Report on the Economic Well-Being of U.S. Households." The report from 2019-May 2020 [3] shows that the percentage of Americans who are not able to pay an unexpected $400 emergency expense has been increasing for the past years from 50% in 2013 to 63% in 2019 (before the COVID-19 financial crisis of 2020).

Furthermore, the data from the U.S. Census Bureau from 2020 [4] shows that 66.5% of American households had a yearly income below $99,999 which means that the percentage of households experiencing financial fragility not only reaches households with an income of less than $50,000 (poverty) but also a significant portion of what is considered middle class. What is more surprising is that the same economic well-being report under "Overall Economic Well-Being in 2019", says that in 2019 75% of adults reported to be "doing at least OK," which was an increase from the 62% reached in 2013.

Therefore, some respondents consider that they are doing OK even when they do not have money to pay an unexpected $400 emergency.

One way to explain the contrast in their responses is that they may be able to generate an income, (so they feel they are doing OK), but they may experience trouble managing their personal finances; this is where the term *financial fragility* comes in. Dr. Annamaria Lusardi [5] identified that one of the reasons for financial fragility is the lack of financial literacy. Dr. Lusardi found that financial illiteracy is most predominant in young and less-educated Americans, with a higher impact among Hispanics and African Americans. Unexpectedly, Dr. Lusardi also found that financial illiterate individuals consider themselves well-informed with regard to financial matters.

Similarly, an article by Neal Gabler, titled "The Secret Shame of Middle-Class Americans" [6], coined the term *financial impotence* as a synonym for financial fragility, financial insecurity, or financial distress. He uses this term to describe the financial difficulty experienced by many middle-class Americans who are financially challenged and struggle to pay their bills even when they have a job and what is considered to be a middle-class income. In his article, Mr. Gabler describes how he made several bad financial decisions in his life due to his financial illiteracy. He also mentions how, in contrast with the descriptions made by Dr. Lusardi, financial impotence affects people from all ethnicities and backgrounds, including the college educated.

In addition to these examples, the Millennial and the *Z* generations[1] are facing some unique financial challenges not seen by previous generations. Firstly, the price of college education has been steadily increasing 3% per year (after inflation) for the past five years, according to the National Center for Education Statistics (NCES) [7], while the income has remained stagnant. Hourly wages have only increased by 0.2% per year (after inflation) since the 1970s [8]. Therefore, for current generations going to college, it is more difficult because the price of college tuition is higher, the amount of money that students could save from summer jobs is lower, and the financial support they could get from their families is also lower, resulting in a high number of students taking out loans. According to Forbes, 44.7 million people in the U.S. have student loans, reaching a total of $1.56 trillion in 2020, with an average value of $32,731 [9]. This financial situation has several negative consequences as college graduates often carry high amounts of debt, preventing them from buying houses or creating new businesses as they must allocate a substantial portion of their income to paying for their student loans. Furthermore, they start saving for retirement at an older age and must stay in the work force longer than previous generations.

Is This Textbook About Borrowing Money?

The objective of this textbook is to provide the decision-making tools required for economic analysis. Since individuals and organizations must finance their projects, the financial tools described in this text may support this decision-making process. Additionally, several renowned financial advisors have criticized the educational system for failing to provide the required financial tools.

In many situations, it is difficult to acquire an expensive asset without taking out a loan. For example, buying a car, buying a house, or attending college would be nearly impossible for most people if it was required to pay the full cost upfront. On the other hand, having a high credit dependency for everyday purchases, such as groceries, could be a sign that one's income is too low.

[1] Generation X considers people born from 1965 to 1980. The Millennial generation refers to people born from 1981 to 1996, followed by the Z generation years 1997 to 2012 the Z stands for *zoomer*, as they were the first generation 'zoom' the internet.

Furthermore, as with a healthy diet, there are certain well-known guidelines such as limiting your carbohydrate intake and eating fruits and vegetables. For example, saving money for retirement is a widely accepted principle; however, there are also personal preferences that affect our decision-making. These preferences are particularly relevant when borrowing money. One recognized financial advisor, Dave Ramsey, asserts that one should only purchase a product after having saved enough money to avoid financing; the opposite perspective is expressed by Robert Kiyosaki, who believes one can generate wealth from borrowing. Both advisors are millionaires, and both have financial seminars at which they outline their financial methods.

I am of the belief that it is proper to use financing to purchase assets that (1) may help to generate wealth—for instance, borrowing for business improvement—and (2) would require a long enough time to save that they would be more expensive to wait for than to finance. For example, buying a used car would be cheaper than renting a car for a long period of time. On the other hand, financing should be kept under control and should be avoided when possible, as it frequently increases the purchase cost.

Is It All Lost?

No. By learning about an engineering economy you are becoming financially literate and acquiring financial knowledge is the first step toward financial stability because you can make decisions that have proper financial support. For example, even if you are able to make the monthly payment for a new car it does not mean that it is the best financial decision as cars depreciate rapidly and you could lose a substantial portion of your investment.

In addition, if you are attending college, you are making the correct decision of getting a professional degree. Although the stories of billionaires such as Bill Gates and Mark Zuckerberg, who did not attend college, do seem enticing, 7 out of 10 adults see the financial benefit of having a college education according to a report by the Federal Reserve. The report also describes that for all ethnic backgrounds, 84% of college graduates have an income over $40,000 per year in comparison to 43% of people with only high school-level education or less [3]. Therefore, pursuing a college education, even when acquiring some student loans, is not a bad idea because this investment provides a good rate of return, as will be described in the following chapters.

While the current economic environment has changed, there are still many opportunities available to those with a college degree. In other terms, although we cannot change the economic weather, having an education will allow you to be better prepared to navigate a financial storm.

Furthermore, although current times seem difficult, life has always been uncertain and challenging. Consider the socioeconomic situation from one hundred years ago. The First World War started in 1914, the Spanish flu happened during the 1920s, the Great Depression occurred in 1930, and Second World War started in 1939. Therefore, the twentieth century was particularly challenging.

The present time is also defying; problems such as: climate change, income inequality, and a complex political environment make it difficult and uncertain. Nevertheless, our time presents great advantages; for instance, we live in an era of great technological advantages, humans can live healthier and longer lives, and we have an ever-growing market full of new services and products. Therefore, although it seems that we are living in a complex environment, current generations are better equipped than previous generations to face their own challenges.

Exercise

According to the article by Neal Gabler, he made multiple financial errors due to his lack of financial experience. Create a list of the main errors and what could have he done to prevent them.

Engineering Economy Project Definition

The development of an engineering economy project involves the following nine steps:

1. Problem statement.

   Express in a concise statement, using clear terms, an issue/problem to be addressed or improved.

2. Problem objective.

   Describe a specific, quantifiable amount of improvement considering the performance indicated in the problem statement.

3. Identify alternatives.

   Identify feasible alternatives. Alternatives must be expressed in quantifiable terms such as costs and benefits. Unquantifiable variables that do not modify benefits such as luxury or convenience should not be considered.

4. Gather information.

   Collect information for each alternative such as costs, benefits, and duration.

5. Define decision variables.

   Identify the decision variable that will support the decision objective and problem statement in engineering economy. There are different variables such as rate of return or an amount of money.

6. Define the selection criterium.

   Define how alternatives would be rated and how the best alternative will be selected.

7. Build the solution model.

   Develop the framework to assess alternatives.

8. Calculate and select the best alternative.

   Use the framework developed in step 7 to calculate the values for each alternative and to determine the best alternative.

9. Audit the result.

   Verify the previous steps and confirm that the best alternative has been selected.

   Let us consider the following example to clarify the previous notation.

Example 1.1

An entrepreneur would like to invest in a new coffee shop at a local university. He would like to know which location, if any, should be selected to generate a return on his investment of at least 10% per year. The university has three locations, one location is near the engineering building, a second location is inside the recreational fitness center, and a third location is near the dormitories. The rent for each location will be affected by transit and space. Consequently, an area with more transit and more space will be more expensive. In addition, the investor is aware that during the vacation periods the number of students on campus decreases.

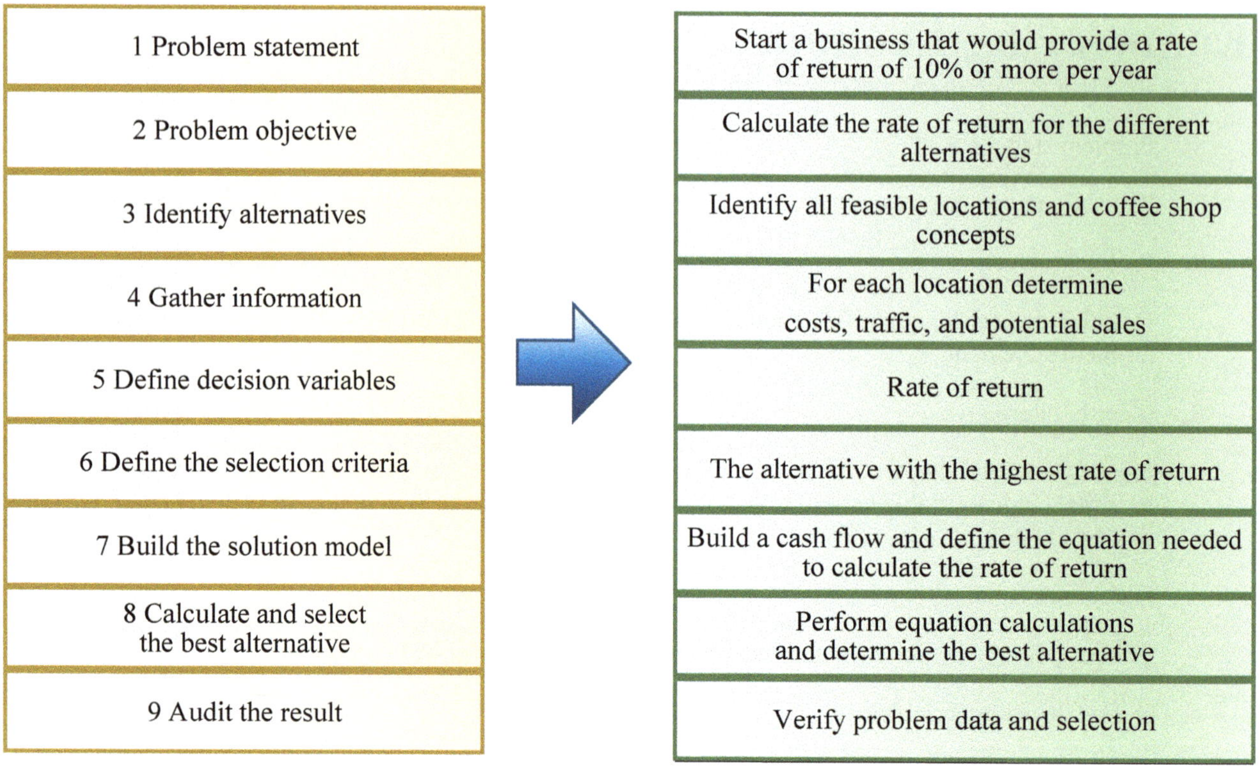

Fig. 1.1. Engineering economy project steps and its application to Example 1.1.

Exercise

Consider that you and your friends (all civil engineering students) would like to open a small consulting firm that offer services to the local community such as making AutoCAD drawings or providing support for house renovation projects. Apply the first six steps of an engineering economy project to describe the information that you would gather for your consulting firm.

Cost Estimations

As described in step 4, an engineering economy project requires information to be gathered. Some cost information can be easily determined by requesting it from the vendor or supplier of a product. However, there is some information that is not known and therefore must be assessed.

Different cost accounting techniques such as contingency theory, standard costing, and activity-based costing (ABC) can be used to estimate costs. However, these techniques are often applied after the fact, meaning that the costs are only determined after the project is completed. Unfortunately, most engineering economy projects are performed before the project begins in order to generate the cost estimates. Therefore, other techniques must be used.

Example 1.2

Assume that we want to generate an engineering economy analysis to determine if we should buy a gasoline car or an electric car. Describe some of the information needed to perform a financial analysis.

Some of the variables we would need to know are as follows:

- Vehicle price
- Economic stimulus if available per option
- Cost of home charging installation for the electric car
- Cost per driven mile for each option based on gas/electric charge
- Maintenance schedule and maintenance cost
- Insurance cost per option
- Financing (interest rate)
- Number of miles that will be driven
- Resale value after "n" number of years
- Decommissioning costs
- Other relevant costs due to use, licenses, state inspection fees, etc.

Therefore, when buying a new car, we may easily estimate the gas costs from information provided from the dealer such as miles per gallon and by also considering our current usage and the average gas cost. However, it may be more difficult to determine maintenance costs if we are switching from an old car to a new car or from an economic car to a luxury car.

It is important to note that estimates are not guessed figures and should be based on information such as average cost of gas. Nevertheless, it is important to understand that estimates, as any type of forecast, will be inaccurate, as it is extremely difficult to generate a perfect prediction.

Estimates can be classified as: rough, intermediate, and detailed. Figure 1.3 shows that more detailed estimates are expected to be more accurate but require a greater number of resources, time, and information.

In the case of car selection, we may use a rough estimate to calculate the number of miles driven per week by simply considering the number of miles driven from home to work (both ways) and any other miles traveled on a routine basis. An intermediate analysis would be more detailed, and it may include additional routine trips and non-routine trips performed on a weekly or monthly basis, such as buying groceries or weekend use. Finally, a detailed estimate could be performed by using a GPS device to track all trips made over a length of time.

Once the data is collected, an analysis can be performed based on the estimate made using different techniques. Some common techniques include per-unit cost, itemized cost, and indexing cost.

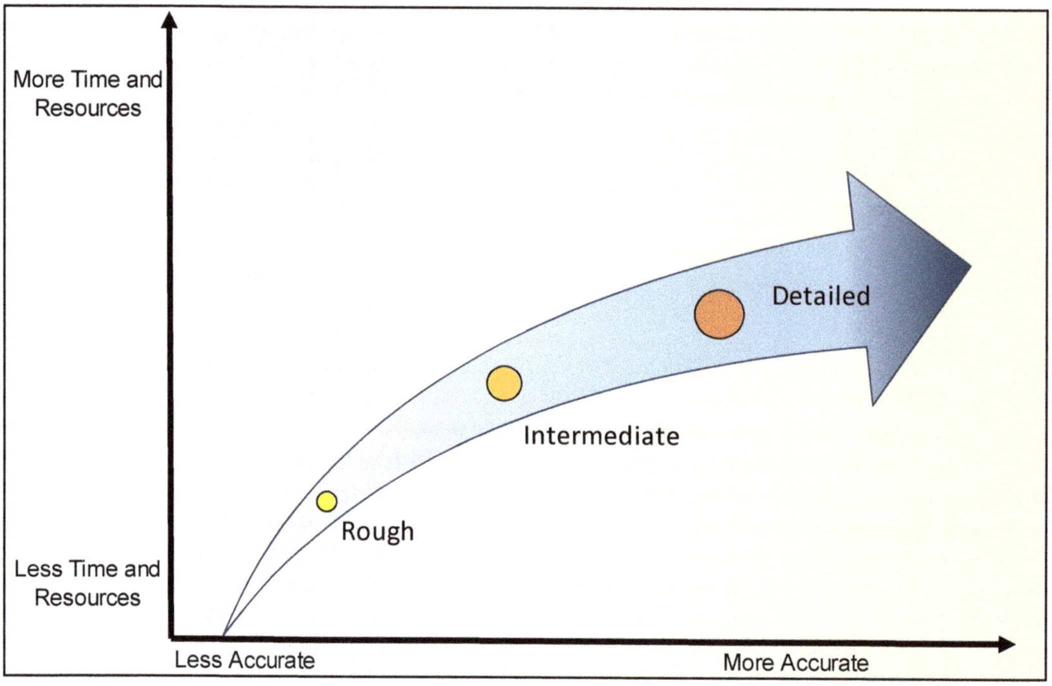

Fig. 1.2. As the number of resources and time increases, it is expected that the estimate will be more detailed and therefore more accurate.

Per-Unit Cost

Estimates the cost based on the cost per unit. For example, the monthly cost for gasoline would be estimated by the number of gallons used per month and the average cost per gallon. Similarly, you may estimate the cost of a new office building by calculating the cost per square foot.

Itemized Cost

Itemized cost considers the different cost elements and then calculates a subtotal value. For example, to determine the monthly cost for a gasoline car we would consider the known monthly costs such as gasoline consumed, average maintenance cost per month, monthly insurance, and so forth. Similarly, in the coffee shop example you may estimate the cost of a cup of coffee by calculating the cost of the different elements needed to prepare a cup such as: water, coffee, cup, and so on.

Indexing

Indexing follows a triangulation calculation method using indexes. To estimate a cost value, you would need a known cost and the indexes for the estimated cost and the known cost. The estimated cost formula can be described as

$$\frac{Estimated\ cost}{Known\ cost} = \frac{Index\ for\ estimated\ cost}{Index\ for\ known\ cost}$$

In other words,

$$Estimated\ cost = \frac{Known\ Cost * Index\ for\ estimated\ cost}{Index\ for\ known\ cost}$$

Example 1.3

Estimate today's oil change cost. Consider that the price of oil change cost is directly affected by the value of a barrel of crude oil. If the price of a barrel of crude oil is $45 today and if the price of a barrel of crude was $95 10 years ago, and the cost of an oil change was $35.

Applying these numbers in the formula we obtain the following:

$$Estimated\ cost = \frac{\$35 * \$45}{\$95} = \$16.57$$

Exercise

Estimate the construction cost for a small house 20 years ago. Consider that a house currently requires an investment of $30 of lumber per square foot. The price of lumber is tied to the Lumber and Wood Products Index. The index has a value today of $476 and 20 years ago had a value of $165.

Answer: $10.39

In addition to its estimation, costs can also be classified based on different categories; for instance, some costs occur at regular intervals and have fixed amounts, such as leases. Additionally, some costs are business related in nature, such as buying supplies, while others are not, such as maintenance. Table 1.1 provides descriptions of different cost categories, with examples.

Table 1.1: Different Cost Categories and Their Application to Example 1.1

	Periodicity		Amount Variability	
	Recurrent	Non-recurrent	Fixed	Variable
Definition	Occur at regular intervals	Occur at irregular intervals	Has a constant value	Has a variable value
Coffee Shop Example	Rent, salaries, licenses	Fines, renovations	Licenses, fees	Supplies, utilities

	Business Nature		Opportunity	
	Internal	External	Sunk	Opportunity
Definition	Related to the purpose and operation of the business	Not directly related to the business purpose	Cost made on previous decisions	Cost of loss opportunity
Coffee Shop Example	Supplies, utilities	Disposal and discharging	Past cost from buying equipment	Missing the opportunity from not selecting an option

	Production		Tax related
	Average	Incremental	Depreciation
Definition	Sum of all costs divided by the number of sold units	Increased unit cost by additional units	The loss of value for a fixed asset due to its use or obsolescence
Coffee Shop Example	All costs divided by the number of units sold	Overtime, additional supply cost.	Coffee maker machine depreciation

Ethics in Engineering Economy

A phrase popularized by Spider-Man, written by Stan Lee, says "with great power comes great responsibility" [10]. This idea dates back to the French Revolution in 1793. The concepts described in this text will allow you to formulate and apply financial concepts such as compound interest. Although we have not yet covered the concept of compound interest, this concept is so powerful that it was described by Albert Einstein as "*the most powerful force in the universe.*" When a genius physicist makes such an assertion, one must pay attention.

Unfortunately, there are multiple examples where individuals and corporations have used their financial knowledge to generate unethical financial gains. As engineers we are entrusted with the development, design, manufacture, use, and operation of the devices and equipment that directly impact the life and well-being of the public. Therefore, as described by the National Society of Professional Engineers (NSPE), engineers should adhere to the "highest standards of honesty and integrity" [12]. The NSPE defines professional ethics as "a set of standards defined by the professional community which provides a guide"[12]. Moreover, as described by the NSPE, the services provided by engineers require honesty, impartiality, fairness, and equity and must be dedicated to the protection of the public health, safety, and welfare.

The description of NSPE code of ethics can be found in the NSPE website (https://www.nspe.org/resources/ethics/code-ethics). It is expected that you will read the entire description of the code and adhere to its description in addition to reviewing the code of ethics from your individual engineering discipline.

The following two examples clarify the application of ethics with respect to engineering economy.

Example 1.4

Fig. 1.3. Example 1.4 job offer sign.

The following sign was posted on a street pole. Assume that a company owned by an engineer posted this sign. Following the NSPE code of ethics, describe whether the sign adheres to the NSPE code of ethics.

This sign advertises a salary of $20,000 per month if you become a real estate investor trainee. According to the sign description all you have to do is to text your contact information to the phone number described and enroll in their trainee program.

For starters, there is nothing wrong with paying a high salary or advertising one. The sign also does not seem to affect the safety, health, and welfare of the public, as the sign itself is safe and could have been posted by an engineer or a company who has competence and is real estate licensed.

However, it is likely that this ad falls under false advertisement, as it seems improbable that a company would pay a trainee a salary of $240,000 a year.

Therefore, if this is the case, the sign would violate the code of ethics by issuing an untruthful public statement. Furthermore, it would violate the point of avoiding deceptive acts, and it would fail to follow ethical and lawful standards.

Example 1.5

Consider a payday loan company who advertises on the internet that the annual percent rate (APR) for a personal loan could be between 4.99% and 450%. They clarify their loan structure by providing a "typical loan example" using an APR of 10% and a fee of 3%.

It is important to question if this advertising is truthful or deceptive as the typical loan example may not be so typical, especially since this type of credit is targeting low customers who are likely to have low FICO scores, and therefore the APR might be far higher than 10%. FICO scores are described in chapter 5 and refer to a person's credit worthiness.

On the other hand, it is important to note that although some operations may be legal, they may be unethical with respect to the welfare of the public. Some of these businesses' practices overcharge interest amounts to low-income individuals who are in a precarious financial situation. Again, it is licit to generate an income, but it is unethical to take advantage of a person's vulnerability to obtain a benefit that goes beyond any reasonable amount.

What Did We Learn in This Chapter?

- Engineering economy is a decision-making tool, providing financial techniques to formulate, calculate, and solve financial problems for professional applications and personal applications.
- Engineering economy projects should be applied to financial problems with intermediate difficulty such as loan payments or rate-of-return problems.
- Estimations are inaccurate by nature but may increase in accuracy based on the amount of information and time provided.
- Costs can be classified by their periodicity, amount variability, business nature, opportunity, production, and tax-related cost.
- Engineering economy is also affected by ethics. Engineers should consider an ethical behavior when performing financial analysis.

Review Questions

Directions: Refer to what you learned in this chapter in order to answer the following questions:

1. Why do engineers need to learn about an engineering economy?
2. How does an engineering economy support the decision-making process?
3. What financial challenges do Millennial and *Z* generations face?
4. Is it always a bad idea to apply for a loan?
5. Are there financial disadvantages to buying a new smartphone every year?
6. Do we need to conduct engineering economy projects for all financial decisions, even the low-complexity problems?
7. Why is it necessary to generate an estimate for engineering economy problems?
8. What is the advantage of correctly classifying costs?
9. Why do we need to consider ethics in engineering economy?

Chapter 1 Practice Problems

Use the following problems to practice your newly acquired knowledge. Answers for even problems are in the back of the book.

1. The chapter's first reading describes the case of Paula, a recent engineering graduate. Describe what financial difficulties do you think most students face at the time of graduation?

2. Use a search engine to research the interest rates offered for student loans.

3. Use a search engine and find the current salaries for your engineering discipline in your current location and two other locations in the city or state where you would like to move.

4. Use a website such as https://neuvoo.com/tax-calculator/ to identify a person's net yearly income (after taxes) if this person receives a yearly income of $110,000 and lives in New York and compare it to if that person moved to Texas. Identify the income differences.

5. Classify the following problems based on level of difficulty: simple, intermediate, and complex:

   a. Deciding which retirement fund will deliver the best results
   b. Selecting between different alternatives on how to buy/pay for a house
   c. Selecting between different credit card alternatives
   d. Selecting your cable/internet provider
   e. Opening a new branch for a manufacturing company
   f. Long-term investments
   g. Planning your retirement overseas
   h. Paying a streaming service per year instead of per month.

6. Which one of the following problems is most suitable for an engineering economic analysis?

   a. One of your two favorite sandwich shops offers a 10-punch loyalty card, and the other does not. Where should you stop today?

   b. A woman has $150,000 in a checking account that pays 2% yearly interest. She can either invest it immediately at a 4% yearly interest rate or wait a week so she obtains an interest rate of 4.15%.

   c. A man needs to define if he wants to lease or to finance a car.

7. Write a list of five personal costs you have incurred in the past week and classify them by periodicity and variability.

8. Determine the estimated cost of a new car today if a similar car had a price of $25,000 4 years ago. Consider that the cost of the car depends on the price of steel. Assume that in this timespan, the price of a ton of steel increased from $850 to $1,100.

9. An American entrepreneur recently traveled to Central America and found an investment opportunity by creating a coffee company that sells Guatemalan and Costa Rican coffee in the U.S. Use the engineering economy project definition to develop the different steps/ activities for this project.

1 Problem statement
2 Problem objective
3 Identify alternatives
4 Gather information
4 Define decision variables
5 Define the selection criteria

10. Classify the following costs based on periodicity, amount variability, business nature, opportunity, production, and tax for the company in the previous.

	Periodicity	Amount variability	Business nature	Opportunity	Production	Tax
Management salary						
Coffee with a discount over volume						
State safety inspection and operation permit						
Fine and lawyer cost						
Buying a delivery truck every 3 years						
Developed an expansion analysis 2 years ago						

11. An engineer developed a training system using virtual reality. During preliminary testing he learned that the system works well for most people, but for some participants from a specific ethnic background it does not work correctly (because of their given physical characteristics). Would it be ethical to include a picture that includes people from this ethnic background in the marketing campaign?

12. In 2015, a car manufacturer experienced an emissions scandal that was the result of a deliberate intent to violate the Clean Air Act. According to multiple reports, it was discovered that this manufacturer diesel cars that were intentionally programmed to activate control their emissions during laboratory emissions testing. About 11 million cars worldwide, including 500,000 in the United States, in model years 2009 through 2015 were affected. Consider this information and the engineering code of ethics and identify which principles could have been violated.

References

[1] B. Chernev. "What Percentage of Startups Fail?" Review42. https://review42.com/what-percentage-of-startups-fail/ (accessed Jun. 28, 2020).

[2] L. Steinberg. "5 Reasons Why 80% Of Retired NFL Players Go Broke," Forbes. https://www.forbes.com/sites/leighsteinberg/2015/02/09/5-reasons-why-80-of-retired-nfl-players-go-broke/ (accessed Feb. 09, 2022).

[3] Board of Governors of the Federal Reserve System. "The Fed - Report on the Economic Well-Being of U.S. Households (SHED)." https://www.federalreserve.gov/publications/report-economic-well-being-us-households.htm (accessed Jun. 28, 2020).

[4] U.S. Census Bureau. "Income and Poverty in the United States: 2020" https://www.census.gov/library/publications/2021/demo/p60-273.html (accessed Feb. 11, 2022).

[5] A. Lusardi and O. S. Mitchell. "Financial Literacy and Retirement Planning in the United States." National Bureau of Economic Research, Working Paper 17108, Jun. 2011. doi: 10.3386/w17108.

[6] H. Kretschemer. "The Secret Shame of Middle-Class Americans," The Atlantic. https://www.theatlantic.com/magazine/archive/2016/05/my-secret-shame/476415/

[7] National Center for Education Statistics "Fast Facts: Tuition Costs of Colleges and Universities (76)." https://nces.ed.gov/fastfacts/display.asp?id=76 (accessed Jun. 28, 2020).

[8] J. Shambaugh and R. Nunn. "Why Wages Aren't Growing in America." Harvard Business Review, Oct. 24, 2017. Accessed: Jun. 29, 2020. [Online]. Available: https://hbr.org/2017/10/why-wages-arent-growing-in-america

[9] Z. Friedman. "Student Loan Debt Statistics In 2020: A Record $1.6 Trillion." Forbes. https://www.forbes.com/sites/zackfriedman/2020/02/03/student-loan-debt-statistics/ (accessed Jan. 05, 2021).

[10] C. Applegren. "With Great Power, Comes Great Responsibility - The Malta Independent." https://www.independent.com.mt/articles/2019-04-09/blogs-opinions/With-great-power-comes-great-responsibility-6736206375 (accessed Jan. 07, 2021).

[11] Jesse Moss. "Dirty Money." https://www.netflix.com/title/80118100 (accessed Jan. 07, 2021).

[12] National Society of Professional Engineers. "Code of Ethics" https://www.nspe.org/resources/ethics/code-ethics (accessed Jul. 13, 2020).

Chapter 2. The Time Value of Money and the Introduction of Present Value, Future Value, and Interest Rates

Paula's Purchasing Power Dilemma

IMG 2.1

As described in the previous chapter Paula had a nice job offer, and she intended to get started on paying her student debt and was working toward generating a monthly budget.

On the way back home, she stopped by the supermarket to buy some groceries and observed that several items had a price increase. Therefore, Paula wondered if she had to consider the price increase when creating the budget. Furthermore, she realized that it wasn't just groceries that were more expensive but so were other items such as gas, and the internet.

To make matters more confusing, after comparing different items she noted that they suffered different percentage increases even when they were related. For instance, the price of milk increased at a lower rate than meat.

This made her wonder how her income would be affected by inflation. Paula's contract stipulated that her salary would be increased by 3% each year. This increase seemed OK considering that in past years the yearly inflation rate was about 2%. However, if inflation for the current year is higher than 7% would this mean that this year her income will be less than in the previous year? Furthermore, if inflation continues to be high what would be the effect of inflation on her student loans and the economy?

Reading Activity: Responding to Paula's Car Story

- If Paula's thinking is correct, what is the effect of inflation in her salary and her future earnings?
- Assume that Paula has an estimate for the next year's inflation rate; how could she use this information to negotiate her salary increase as well as her rent's lease?
- If Paula has an accurate inflation estimate for the next 10 years, could she use it to determine the future price for her rent?
- Would knowing the inflation estimate for the past 10 years be useful? for example, would it be useful to know what the price of an item was 10 years ago?

Learning Outcomes

After covering this chapter, the student should be able to do the following:

1. Learn the relevance of the time value of money.
2. Learn how to calculate an interest amount and an interest rate.
3. Learn about simple and compound interest.
4. Learn how to calculate nominal and effective interest rates including continuous compounding.

Introduction: The Time Value of Money

As described in the reading, items are worth different amounts depending on time. This is the single most important concept in an engineering economy. The concept is called economical equivalence and can be described as two *different* amounts of money placed at different time periods are *equivalent* if they have the same economic power.

To explain this concept let us consider the following example.

Example 2.1

Consider that a gallon of milk has a current cost of $3.00 and suppose that the inflation rate per year is 7%. Assume that the sale price of a gallon of milk is only affected by inflation and no other factor. What will the financial equivalence needed to buy a gallon of milk be one year from today?

$$\$3.00 * 1.07 = 3.21$$

In this example we multiplied by 1.0 because we wanted to maintain the full value of the item, $3.00, and we added the increment of 0.07 (or 7%) as this increment is the change in value for the item given by the yearly inflation.

Although this example is simple it shows the following:

- Even when two quantities are different, and they are placed at different time periods, they have equivalent economic power.
- The rate of equivalence is expressed as a rate, and this rate is described as an interest rate.

The rate of equivalence is also known as the interest rate and is used to determine the equivalence required for using someone else's money. For example, if you use a credit card you will be charged an interest rate. On the contrary, if you are an investor, this rate of equivalence is called rate of return (ROR) and is charged to someone for borrowing money. Therefore, a bank (an investor) is paid a rate of return for lending money.

Example 2.2 Equivalence

John's friend is selling his used car for $15,000. Mark, John's friend, wants to like to buy John's car as it is in very good condition, but he does not have the money. John gives Mark three different financing alternatives: (a) pay the full amount today, (b) pay $500 each month for 36 months, or (c) pay $5,000 at the end of each year plus the interest generated each year at a 6% yearly rate.

Example 2.2 Equivalence (Continued)

Let us review each option:

a) Pay the full amount today, \$15,000, because it is the principal amount, with no interest.

b) Pay \$500 each month for 36 months.

\$500 * 36 months = \$18,000

Since the principal (value of car today) is \$15,000 the interest amount generated is \$3,000.

c) Pay 5,000 at the end of each year and pay the interest generated by the principal each year using a 6% yearly rate.

Table 2.1: Example 2.2 Equivalence

Period (n)	Principal at Start of the Period	Interest Amount i_A	End-of-Year Balance	Interest Payment	Principal Payment	Total Payment
0	\$15,000	\$900	\$15,900	\$900	\$5,000	\$5,900
1	\$15,900 - \$5,900 = \$10,000	\$600	\$10,600	\$600	\$5,000	\$5,600
2	\$10,600 - \$5,600 = \$5,000	\$300	\$10,300	\$300	\$5,000	\$5,300
				Total payment		\$16,800

Therefore, to calculate the interest amount for the third option we subtract the total payment from the principal and calculate the interest amount i_A:

\$16,800 - \$15,000 = \$1,800

According to these calculations, the three options show different interest amounts. The first option has the lowest financial cost as the interest amount i_A = \$0, while the second option pays a higher interest amount, i_A = \$3,000.

Nevertheless, it should be noted that although these plans have different *interest amounts*, the plans are *financially equivalent*.

Introduction to Interest Amount and Interest Rates

To calculate the interest amount paid or earned and the interest rate, we need to consider the following notations:

P = Principal (amount borrowed or invested at present time)

i_A = Interest amount (amount paid from borrowing or earned from lending)

i = Interest rate

n = Number of periods

F = Future (amount in a future time in the future)

To clarify the previous notations, let us consider the following example.

Example 2.3

George plays in a band, and he needs a new guitar. The guitar he picked has a price of $500, and the store is offering him the store's credit card to buy the guitar. According to the agreement George will make no payments during the year and will pay $550 to the store one year after the date of purchase.

Determine the interest amount paid and the interest rate.

The principal amount is the amount of money George borrows. The principal is represented by the letter P. For this example, it is $500.

The interest amount paid is i_A and is calculated by subtracting the principal "P" from the amount from the future F. The amount from the future is the amount of money that George will pay one year from today. In this example it is equal to $550.

Finally, the number of time periods n is 1 because the loan is for one year. Therefore,

$$i_A = F - P$$

$$i_A = \$550 - \$500 = \$50$$

The interest rate i calculated for the period of one year is.

$$i = \frac{i_A}{P}$$

$$i = \frac{\$50}{\$500} = 0.1 \text{ or } 10\%$$

In this example, the number of periods is one year. However, there could be multiple periods within a year such as months, weeks, or even days. On the other end, there could be an interest rate for periods longer than one year. The following examples will clarify how to handle different time units.

Simple Interest

When calculating the interest amount using simple interest for one period, we use the following formula:

Equation 2.1: Interest amount formula for a single period

$$i_A = P * i$$

Applying this equation to Problem 1.2 we obtain the following:

$$i_A = \$500 * 10\% = \$50$$

When calculating simple interest for different periods, we apply the same interest amount for each period, resulting in the following formula:

Equation 2.2:Interest amount formula for multiple periods

$$i_A = P * i * n$$

where "n" is the number of periods. To clarify this concept, consider the following example.

Example 2.4

Angela will pay a $100 loan after four years and she will be charged a 10% per year in simple interest. Determine the interest amount that will be paid in this loan.

In this case we use the formula for multiple periods and obtain the following:

$$i_A = P * i * n$$

$$i_A = \$100 * 0.1 * 4 = \$100 * 0.1 * 4 = \$40$$

It is important to note that that the principal amount for Equation 2.2 remains constant. The following example, Example 2.5, describes a situation where a loan is paid with multiple payments.

Example 2.5

Matthew is borrowing $1,000 today. He is charged using simple interest, at a yearly rate of 10%. Determine his payment amount if he wants to split his payment into four equal payments that include the principal payment and its corresponding interest amount.

Considering Equation 2.1

$$i_A = P * i$$

$$i_A = \$1{,}000 * 10\% = \$1{,}000 * 0.1 = \$100$$

The principal amount per period is as follows:

$$\frac{P}{4} = \frac{\$1{,}000}{4} = \$250$$

Considering Equation 2.1 we calculate the interest rate for one period. Note that all periods pay the same principal amount.

$$i_A = P * i$$

$$i_A = \$250 * 0.10 * = \$25$$

Therefore, the interest amount for all 4 periods is

$$i_A = P * i * n$$

$$i_A = \$250 * 0.10 * 4 = \$100$$

Exercise

Determine the interest amount that you would pay after borrowing $50 for 10 periods. Use simple interest and a 5% interest rate per period.

Answer: $25

Compound Interest

FIG 2.1. Benjamin Franklin

Benjamin Franklin described compound interest as "money makes money. And the money that money makes, makes money" [1]. Compound interest is paid on both the principal and on the accumulated interest.

To calculate compound interest, the principal amount is re-calculated for each period by adding the principal and the interest amount generated by that period. In turn, the re-calculated principal amount is used to determine the interest amount.

In the case of a positive interest rate the principal amount for a future period is higher than one of the preceding period. This process is called capitalization. Let us clarify this description using the following example.

Example 2.6

For her 10th birthday, Alyson received $5,000 from her grandfather. She invested this amount into a savings account that paid 5% per year. If she does not withdraw any money from her savings, determine how much interest will be generated after 5 years. Perform this calculation using compound interest.

To calculate the interest amount, we calculate the principal at the beginning of each period and then multiply that by the interest rate per period.

Table 2.2. Example 2.6

Period (n)	Principal at the Start of the Period	Interest Rate	Interest Amount i_A at the end of the n period
0	$5,000.00	5.00%	$250.00
1	$5,000.00 + $250.00 = $5,250.00	5.00%	$262.50
2	$5,250.00 + $262.50 = $5,512.50	5.00%	$275.63
3	$5,512.50 + $275.63 = $5,788.13	5.00%	$289.41
4	$5,788.13 + $289.41 = $6,077.53	5.00%	$303.88
5	$6,077.53 + $303.88 = $6,381.41		

Let us review each column. The first value is the period value, n. According to the description the starting point is Alyson's 10th birthday; however, the period value in the table describes it as period 0 instead of period 10. Because the calculation does not consider Alyson's age, but it is instead performed according to her investment.

Therefore, the period is 0, as the investment just occurred; after exactly one year the period is 1, and so on. Thus, the period value follows the same logic as birthday logic, where an individual's birthday occurs after a full period has been completed.

As we observe, the interest amount for the following periods increases as the principal at the start of the period. This is done by adding the updated principal value and the interest amount for each period. In addition, it is important to note that there is no interest amount calculation for year 5 because the investment is not sustained from year 5 to year 6. Refer to Fig. 2.2 1 to clarify this point.

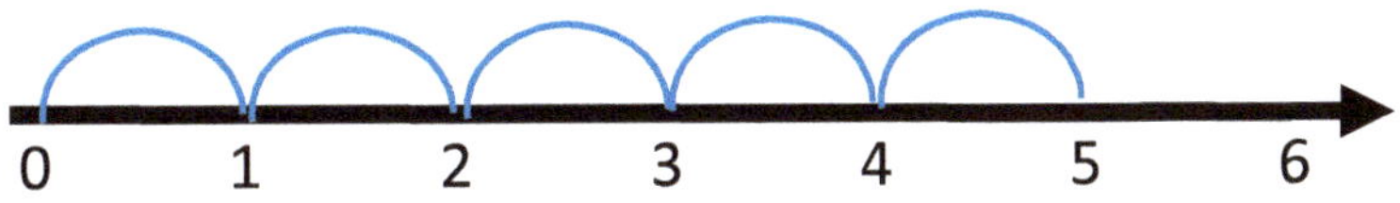

Fig. 2.2. Example 2.5: Alyson's birthday periods.

Exercise
Determine the interest amount that you would pay after borrowing $50 for four periods. Use compound interest and a 5% interest rate. Answer: $60.78

Future from a Present Calculation

The previous example shows a way to calculate an interest amount for each of the different periods. However, following this methodology would become cumbersome and lengthy for situations with a large number of periods. Therefore, we will develop a mathematical formula to perform these complicated calculations.

Consider the calculation of the principal at the start of the period for period 1. This amount was calculated by multiplying the principal from period 0 and then by adding the interest amount from the first period. This could be expressed by the following formula:

Equation 2.3: Future for a single period

$$F_1 = P_0(1 + i)$$

Where:

F_1 = Future value for period 1

P_0 = Present value for period 0

i = interest rate

Similarly, for period 2 we use the same formula and obtain.

Equation 2.4: Future for period 2 from a present value in period 1 for a single period

$$F_2 = P_1(1 + i)$$

Where:

F_2 = future value for period 2

P_1 = present value for period 1, which is equal to F_1 from Equation 2.3

i = interest rate

On the other hand, since $F_1 = P_1$ and $F_1 = P_0(1 + i)$ then we substitute P_1 into Equation 2.4 and obtain the following:

Equation 2.5: Future for year 2 using values from equation 3

$$F_2 = P_0(1 + i) * (1 + i)$$

$$F_2 = P_0(1 + i)^2$$

Therefore, we could continue with this sequence for all of the following periods, obtain the following equation:

Equation 2.6: Future for period n from a present using compound interest

$$F = P(1 + i)^n$$

Where:

F = future value for period n

P = present value

i = interest rate

Let us apply this formula in the following situation. School Street is a significant street in downtown Boston, Massachusetts. This street is only a couple of blocks long and is named for being the place of the first public school in the United States.

School Street is also home to one of the first branches of BankFive, formerly known as The Boston Five Cents Savings Bank. The bank was chartered on April 10, 1855, and this name was intended to attract low-income populations as the minimum deposit required to open an account was five cents, instead of one dollar as other banks at that time [2]. Assume that the average inflation rate in the United States is 2.08% per year between 1855 and 2020 [3].

Fig. 2.3. The Boston Five Cents Savings Bank building located in School St. in downtown Boston MA.

Example 2.7

Can we determine the future value of 5 cents from 1855 in 2020's money?

To calculate this value, first we determine the number of years between 2020 and 1855. Therefore, n= 165. we then calculate present values year by year as we did in Example 2.5. Therefore, the value for 1856 is the principal value for 1855, multiplied by 1, plus the interest rate:

$$\$0.05 * (1+0.0208) = 0.05104$$

We could then continue these calculations as described in Table 2.3.

Table 2.2: Example 2.7 Future Value Calculation

Year	Principal at the Start of the Period	Interest rate	Interest Amount (i_A)
1855	$0.05000	2.08%	$0.00104
1856	$0.05104	2.08%	$0.00106
1857	$0.05210	2.08%	$0.00108
⋮			

However, instead of performing 165 yearly calculations, we consider Equation 2.6 to solve this value

$$F = P(1 + i)^n$$

where

F = future value for period 165 (year 2020)

P = present value for period 0 (year 1885)

i = interest rate (assuming 2.08% yearly inflation)

Example 2.7 (Continued)

Thus, the five-cent problem is solved as:

$$F = P * (1 * i)^n$$

$$F_{165} = \$0.05\ (1 * \ 0.0208)^{165}$$

$$F_{165} = \ \$0.05 * \ 29.8682 \ = \ \$1.4934$$

Consequently, a person would be able to open a bank account with $1.4934 today dollars.

Exercise

Determine how much one dollar in 1855 would be worth in todays. Assume an interest rate of 2.08% per year.

Answer: $29.86821

For simple interest the Future formula is defined by Equation

Equation 2.7 Future for period n from a present considering simple interest

$$F = P + (P * i * n)$$

F = Future (amount in a future time in the future)

P = Present (amount in a future time in the present)

i = Interest rate

n = Number of periods

Example 2.8 determines the number of periods required to reach a certain future value using simple and compound interest.

Example 2.8

Stephen invested $1,000 in a friend's business. His friend's business pays 10% interest rate per period (each period is one year long). Determine how many periods are required to duplicate his investment (a) using simple interest and (b) using compound interest.

We consider Equation 2.7 for part (a) and Equation 2.6 for part (b)

(a) $P = \$1{,}000$

$F = \$2{,}000$

$i = 10\ \%$

Example 2.8 (Continued)

$$F = P + (P * i * n)$$

$$\$2{,}000 = \$1{,}000 + (\$1{,}000 * 0.1 * n)$$

$$n = (\$2{,}000 - \$1{,}000)/(\$1{,}000 * 0.1)$$

$$n = 10\, periods$$

(b) We consider the Future formula for compound interest $F = P(1+i)^n$

$$\$2{,}000 = \$1{,}000(1+0.1)^n$$

$$\frac{\$2{,}000}{\$1{,}000} = (1.1)^n$$

$$\ln(2) = \mathrm{n} * \ln(1.1)$$

$$0.6931 = n * 0.0953$$

$$n = \frac{0.6931}{0.0953} = 7.2728 \approx 7.3\, periods$$

Present from a Future Calculation

Having generated Equation 2.6 to calculate a future value from present, we can calculate a present value from a future value:

The following Equation 2.8 shows how to calculate a Present value from a Future value.

Equation 2.8: Present value from a future value.

$$P = \frac{F}{(1+i)^n}$$

Where:

F = future value for period n

P = present value

i = interest rate

n = Number of periods

Example 2.9

According to U.S. news the average college tuition for year 2020-2021 is $35,087 [4], assuming the cost of college steadily increased by 3% for the past five past years. Determine the cost of college five years ago.

$$P = \frac{F}{(1+i)^n}$$

$F = \$35{,}087$

$i = 3\%$

$n = 5$

$$P = \frac{\$35{,}087}{(1+0.03)^5} = \frac{\$35{,}087}{(1.1592)} = \$30{,}268.28$$

Therefore, the tuition cost five years ago was $4,818.71lower.

Exercise

Assume that the yearly inflation in the United States was 1.75% per year between 2015 and 2020 [5]. Determine the price of a car in 2015 if the selling price in 2020 is $22,500.

Answer: $20,630.53

Drawing Cash Flows

One important step in understanding engineering economy problems is to draw cash flows and to understand the sequence and the perspective of the problem. Cash flows represent inflows and outflows of money within a timeline. As observed in Fig. 2.4 a cash flow is represented by the following components. A horizontal line depicts the time flow and acts as the base of the cash flow diagram. The number of periods considered are represented by the numbers, starting with the present as time = 0 and ending at the final period as time n.

Arrows are used to depict the direction of cash flows. The upward arrow in time 0 represents a positive inflow. A positive inflow is when a person or a business entity receives money. Money can be received for different reasons, for example, when a person or business entity receives payments, gifts, loans, or tax returns.

The negative outflow in year 3 could represent an expense, a payment, or even an investment.

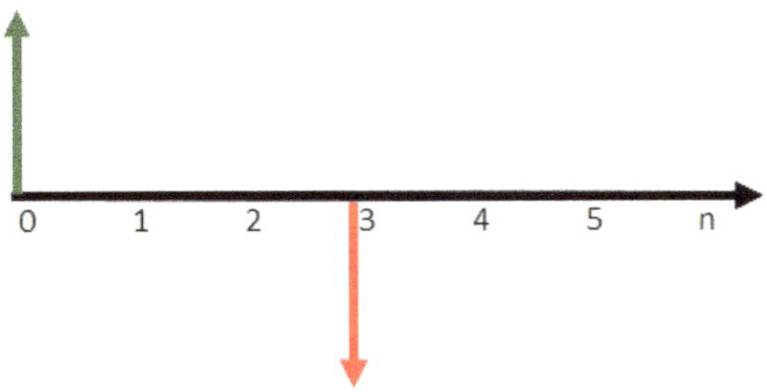

Fig. 2.4. Cash flow.

In addition to considering inflows and outflows of cash, it is important to consider the perspective of the cash flow. This is explained in Example 2.10.

Example 2.10

Figure 2.5 shows Example 2.10 from the buyer's perspective. In this flow we should note that the inflow arrow is longer than the outflow arrow. Although these arrows do not follow a perfect scale, it is helpful to indicate that the amount of money borrowed is greater than the amount in each payment. In addition, although the buyer did not receive an amount of money, they received a car for the value of the loan.

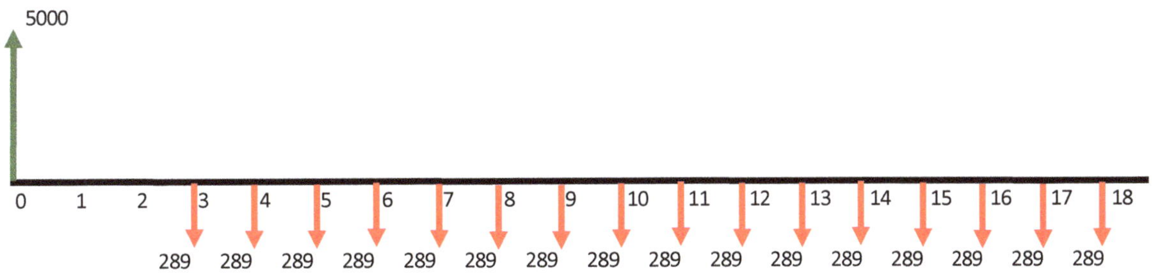

Fig. 2.5. Example 2.10 buyer's perspective

Figure 2.6 shows the cash flow from the seller's perspective. Here, in time 0, the seller has an outflow as they are lending the car with a value of $5,000. Every time the buyer makes a payment it is an outflow for the buyer but is an inflow for the seller.

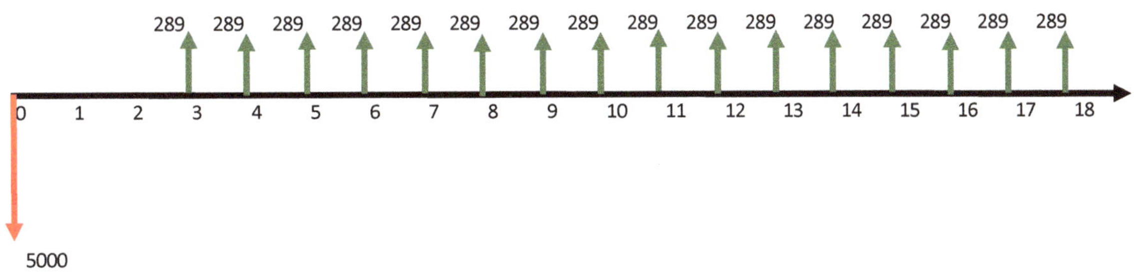

Fig. 2.6. Example 2.10 Seller's perspective

It is important to note that the financial calculations for Example 2.10 would have the same numerical amount, but the "sign" would be opposite for each perspective. We will go over this in the following chapters.

Exercise

Draw the following cash flow. Michelle invested $10,000 in a small business. She received $250 for the first 10 months and then received $450 for months 11 to 20. In month 21 she sold her share in the business for $4,000.

Compounding Periods

In previous examples we considered situations where the time period's frequency unit for the interest rate is the same as the time period's frequency unit for the cash flow. For example, when calculating the future value for a 5-year deposit with a 2% interest rate per year, both time period frequency units are in years. However, this is not always the case as sometimes the interest rate time period frequency unit is different than the time period frequency unit for the cash flow. For example, you could be given a yearly interest rate of 3% with monthly compounding interest. Monthly compounding means that every month the principal amount is recalculated by adding the interest amount to the principal.

Table 2.2 in Example 2.6 shows a situation where the interest amount is added to the principal value at the completion of each year. Consider that the capitalization for the first year occurs after 365 days have been completed. Therefore, the principal at the beginning of the first year is the principal from the previous period plus the interest amount generated after one year.

Bringing Yearly Interest Rates to Periods Shorter Than One Year

In certain situations, a loan is compounded in periods that are shorter than one year. These periods could be by day, week, biweekly (two-week), monthly, quarterly (four-month), or semesterly (six-month).

To convert a yearly period into a shorter period, we simply divide the nominal interest rate per year by the number of periods in a year.

Equation 2.9: Interest rate per period from nominal interest rate

$$i_m = \frac{i_n}{m}$$

where:

i_m = interest rate per period

i_n = nominal interest rate per year

m = number of periods in a year

Example 2.11

A student has a credit card with a yearly interest rate of 18%. Determine the interest rate per month.

$i_m = ?$

$i_n = 18\%$

$m = 12\ periods$

$$i_p = \frac{18\%}{12} = 1.5\%$$

Exercise

Use a yearly interest rate of 18% and determine the interest rate per (a) quarter, (b) semester and (c) week.

Answer (a): 4.5%

Answer (b): 9.0%

Answer (c): 0.3461%

Nominal Interest Rate and Effective Interest Rate

There are situations in which a business describes an interest rate but when the customer sees the monthly statement the calculation does not seem to correspond to the interest rate described in the initial offer. For example, consider that when buying a new car, the car dealer offers a monthly interest rate of 0.5%. Therefore, the customer would quickly assume that the interest rate per year would be 6%. This value is obtained by multiplying the monthly interest rate and the number of months within a year. This rate is known as nominal interest rate, and it is described in Equation 2.10.

Equation 2.10: Nominal interest rate

$$i_n = i_m * m$$

where

i_n = nominal interest rate per year

i_m = interest rate per period

m = number of periods in a year

On the other hand, it is likely that for the described situation when the customer reads the monthly statement the amount charged by the car dealer may be higher than the 6% per year. This could be because the car dealer performs a monthly capitalization. A monthly capitalization would consider the unpaid balance to calculate the interest amount. In other words, the effective interest rate considers that at each period, in this case each month, the principal value is compounded, and this would occur for all m compounding periods within a year. This is calculated by using Equation 2.11.

Equation 2.11: Effective interest rate

$$i_{eff} = \left(1 + \frac{i_n}{m}\right)^m - 1$$

This equation can also be written in the following form:

$$i_{eff} = [(1 + i_m)^m - 1] * 100$$

where

i_{eff} = effective interest rate per year

i_m = interest rate per period

m = number of periods in a year

Example 2.12

Maggie, an engineering student, needed $450 to fix her car. Today she got a short-term loan from a payday company and agreed to pay $500 two weeks from today.

Determine the following:

a) The interest amount that Maggie will pay.

b) The interest rate the Maggie will pay for the two-week period.

c) The nominal yearly rate

d) The effective yearly rate

a) The interest amount i_A is calculated by the difference between the F-P, from Example 2.3:

$$i_A = F - P$$

$$i_A = \$500 - \$450 = \$50$$

b) The interest rate for the period is calculated by

$$i_m = \frac{i_A}{P}$$

$$i_m = \frac{\$50}{\$450} = 0.1111 * 100 = 11.11\%$$

Example 2.12 (continued)

c) The nominal interest rate, since there are 26 periods of two weeks in a year, is calculated using Equation 2.9:

$$i_n = i_m * m$$

$$i_n = 11.11\% * 26 = 288.8889\%$$

d) The effective interest rate is calculated using Equation 2.10:

$$i_{eff} = \left(1 + \frac{i_n}{m}\right)^m - 1$$

$$i_{eff} = \left(1 + \frac{2.8888}{26}\right)^{26} - 1 = [(1 + 0.1111)^{26} - 1]$$

$$i_{eff} = [(1 + 0.1111)^{26} - 1] * 100 = (15.4732 - 1) * 100 = 1447.32\%$$

Using the second form of the formula

$$i_{eff} = [(1 + i_m)^m - 1] * 100$$

$$i_{eff} = [(1 + 0.1111)^{26} - 1] * 100 = (15.4732 - 1) * 100 = 1447.32\%$$

Exercise

Assume that Maggie from Example 2.9 also borrowed \$300 today for additional expenses from a community credit union. She agreed to pay \$400 six months from today. Determine the following:

(a) The interest amount that Maggie will pay.

(b) The interest rate for the six-month period

(c) The nominal yearly rate

(d) The effective yearly rate

Answer (a): \$100 Answer (b): 33.33% Answer (c): 66.66% Answer (d): 77.77%

Continuous Compounding

When compounding periods become shorter the effective interest rate follows continuous compounding. Continuous compounding occurs when the compounding periods are infinitely short and the number of periods in a year becomes infinite.

Therefore, considering the effective interest rate formula and calculating the value limit when m is infinite, we obtain the following:

$$\lim_{m \to \infty} i = \lim_{m \to \infty} \left(1 + \frac{i_n}{m}\right)^m - 1$$

Resulting in

Equation 2.12: Effective interest rate for continuous compounding.

$$i_{eff} = e^{i_n} - 1$$

Example 2.13

A credit card has a promotional rate of 0.5% per month. Determine the effective interest rate if the card interest rate is calculated using continuous compounding, Equation 2.12.

$$i = e^{i_n} - 1$$

Since there are 12 months in a year the nominal interest rate is calculated as:

$$i_n = 0.5\% * 12 = 6\%$$

The effective interest rate for continuous compounding is

$$i_{eff} = e^{i_n} - 1$$

$$i_{eff} = e^{0.06} - 1$$

$$i_{eff} = 1.061836 - 1$$

$$i_{eff} = 0.061836 * 100 = 6.1836\%$$

The effective interest rate with continuous compounding is higher than the nominal interest rate.

Exercise

Michael needs to borrow $500 to fix his computer for the spring semester. He went to a payday loan company with the idea that he would cover the loan with his next paycheck, which is happening in four weeks. The company will require Michael to pay $590 in four weeks.

(a) Determine the interest amount charged for the four-week period.

(b) Determine the interest rate charged for the four-week period.

Exercise

c) Determine the yearly nominal interest rate charged.

d) Determine the effective interest rate charged.

Answer a): $590 - $500 = $90 Answer b): $90 / $500 = 0.18 * 100 = 18% per period

Answer c): Number of periods = 52 weeks per year / 4 weeks per period = 13 periods, Nominal rate = 18% * 13 = 234%

Answer d) i_A= (1+0.18)^13 - 1 = 8.5993 – 1 = 7.5993 * 100 = 759.93% per year

Using Excel to Previous Examples

Microsoft (MS) Excel is a powerful tool to solve financial problems. The following shows how to solve some of the previous examples using Excel, as seen in Fig. 2.7. For novice Excel users a simple definition of a spreadsheet is that it is a collection of cells. In each cell you can enter values, or you can enter your own formulas to perform calculations, as shown in the following examples.

In addition, Excel includes many built-in functions for calculating values such as averages, sums, or standard deviations. To calculate an average, you can simply type in the associated formula and select the range of cells you want to include. If you are unfamiliar with Excel, we recommend that you select the help option from the main menu, where you can find many useful tutorials.

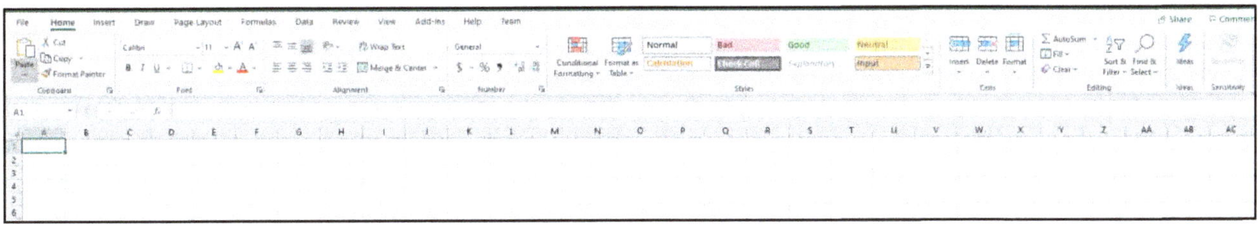

Fig. 2.7. Excel.

Example 2.1: Excel solution

Open Excel, enter each value in a cell, and type the equal sign to enter the formula "=," as seen in Fig. 2.8 below. You can then select the cells and operators for multiplication.

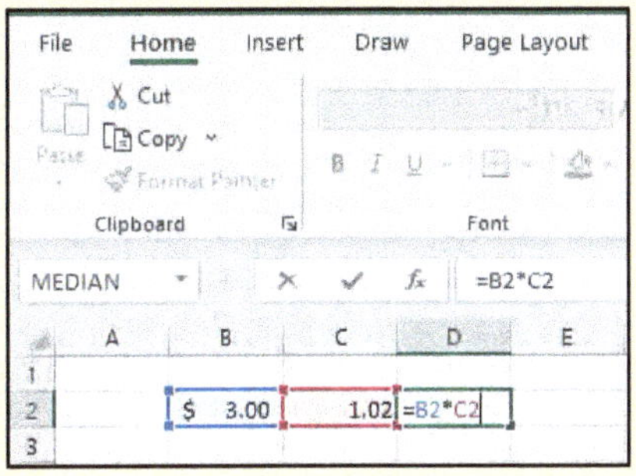

Fig 2.8. Example 2.1 Excel solution.

Figure 2.4 shows a cash flow with two arrows, one representing an input and the other representing an output. In order to have a consistent equation, Excel will calculate the output of a financial equation with a different sign. For example, if you are calculating a future value and you enter a positive value with a positive sign, the future value will have a negative sign and vice versa. This is clearly shown in the following examples.

Example: 2.7 Excel solution

Assume that the average inflation rate in the United States is 2.08% per year between 1855 and 2020 [3]. Could we determine the equivalent value of 5 cents from 1855 in today's money?

To solve this problem, as shown on Fig. 2.9 below, we type "=FV," the future value formula FV, and then we select the required cells by pointing and left-clicking with the computer mouse.

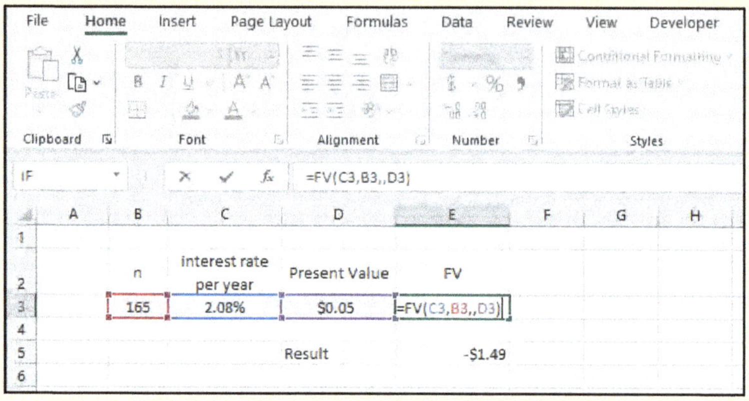

Fig. 2.9. Example 2.7 Excel solution.

Example: 2.9 Excel Solution

According to U.S. news the average college tuition for year 2020 - 2021 is $35,087 [4], assuming the cost of college steadily increased by 3% for the past five past years. Determine the cost of college five years ago. To solve this problem as shown in Fig. 2.10 below, type "=PV," the present value formula PV, and select the required cells.

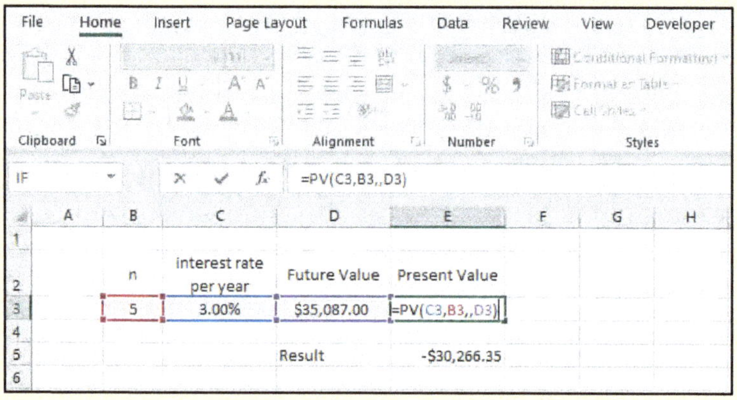

Fig. 2.10. Example 2.9 Excel solution.

What Did We Learn in This Chapter?

- The time value of money, known as the concept of economical equivalence, describes that two different amounts of money can be equivalent if they have the same economic power.
- Interest rates make two different amounts of money equivalent, and we can calculate equivalent amounts for future or past amounts.
- Cashflows represent inflows and outflows of money within a timeline.
- Interest capitalization can be simple or compound. Simple interest does not re-calculate the principal amount for the different periods. On the other hand, compound interest re-calculates the principal amount by adding the interest amount to the principal. The re-calculation is also known as capitalization.
- Nominal interest rate per year is calculated by multiplying the interest rate per period by the number of periods in the year.
- Effective interest rate considers multiple compounding periods within a year.
- Microsoft Excel is a powerful tool to find solutions to financial problems.

Review Questions

Directions: Refer to what you learned in this chapter in order to answer the following questions:

1. Explain how it is possible for two different amounts to be equivalent.
2. What is the difference between interest amount and interest rate?
3. When calculating a future value, why would using compound interest formulas generate higher amounts than simple interest would?
4. If you wanted an investment to grow faster, would you prefer an investment with shorter or longer capitalization periods?
5. Why is it useful to include a cash flow diagram in the solution of an engineering economy problem?
6. How do you define capitalization?
7. If you were to borrow money, would you prefer to pay it using continuous compounding or monthly capitalization?
8. If you were to borrow, would you prefer a charge of 1% per month, or a nominal charge of 12%?
9. How would you describe continuous compounding interest?

Chapter 2 Formulas

Equation 2.13: Interest amount formula for a single period

$$i_A = P * i$$

or given present and future values:

$$i_A = F - P$$

Equation 2.14:Interest amount formula for multiple periods

$$i_A = P * i * n$$

Equation 2.15: Future for period n from a present

$$F = P(1 + i)^n$$

Equation 2.16: Present value from a future value.

$$P = \frac{F}{(1 + i)^n}$$

Equation 2.17: Interest rate per period from nominal interest rate

$$i_m = \frac{i_n}{m}$$

Equation 2.18: Nominal interest rate

$$i_n = i_m * m$$

Equation 2.19: Effective interest rate

$$i_{eff} = \left(1 + \frac{i_n}{m}\right)^m - 1$$

or considering the interest rate per period:

$$i_{eff} = [(1 + i_m)^m - 1] * 100$$

Equation 2.20: Effective interest rate for continuous compounding.

$$i_{eff} = e^{i_n} - 1$$

Chapter 2 Practice Problems

Use the following problems to practice your newly acquired knowledge. Answers for even problems are in the back of the book.

Generate Cash Flows Problems

1. Generate a cash flow for the following description. Jonah wants to retire after 25 years of working in a construction company. He plans to include different investments towards his retirement. First, he plans to include his current savings of $50,000. Second, he plans to include the different fidelity bonuses that his company is offering if he keeps working for the company. The first fidelity bonus is for $5,000 and is given after working for the company for five consecutive years. The second fidelity bonus is $10,000 and is given after working for the company for 10 consecutive years. The third fidelity bonus is for $15,000 and is given after working for the company for 15 consecutive years. The fourth fidelity bonus is $20,000 and is given after working for the company for 20 consecutive years. The fifth and final fidelity bonus is for $30,000 and is given after working for the company for 25 consecutive years.

2. Generate a cash flow for the following description. Jennifer will graduate in May 2024. She is planning to start working in January 2025. Her income per month will be $7,500 for her first year, and her company has agreed to increase her salary by 2.5% after her first year of employment (to cover inflation). Jennifer has monthly living expenses totaling $3,000. In addition, she pays $2,500 in student loans every four months starting in September 2020.

3. On January 1, the owner of a small supermarket borrowed $370,000 to improve his business. In February he spent $50,000 improving the interior and outside lighting. In March he spent $35,000 on new refrigeration equipment. In April he invested $65,000 in self-checkout cash registers. In May he invested in new shopping carts, shelves, and the parking lot for 150,000. He used the remaining money to pay a fixed amount for the months of February to August to his vendors. Generate the cash flow diagram for the money borrowed and his incurred expenses.

Simple and Compound Interest

4. Calculate the interest amount from a loan with a principal amount of $24,000 and an interest rate of 3% per month for 12 months. Use simple interest.

5. Determine how long it would take to triple an investment with a principal value of $4,000, consider an interest rate of 5% per month. Solve using a) simple interest and b) compound interest.

6. Use simple interest to calculate the following. Joe has a $500 investment if his investment pays a rate of return of 2% per period. Calculate the number of periods required to (a) generate an interest amount of $250 and (b) generate an interest amount of $1,000.

7. Use the simple interest formula to determine the interest amount that will be paid every month from borrowing $10,000 for 2 years, use a 15% interest rate per year.

8. Determine the interest amount that will be paid when borrowing $15,000 for 3 years, using a 5% interest rate per year compounded annually.

9. Alex is borrowing $150 and has agreed to pay his debt in 10 payments; however, he needs to select between two interest rates. He could either calculate his payment by using 5% per period using simple interest or pay 3% per period with a compound interest rate.

10. Determine the number of years required for a crypto currency to reach a value of $350,000 if its current value is $25,000 and its growth rate is 12% per year.

Payment Calculation Tables

11. Eddie bought a used car and agreed to make six payments, one payment every month. Use the following table to fill out the total payment per month; consider simple interest and a rate of 3% per month. Eddie will make six equal payments on the principal, not on the total payment, and he will pay the full interest amount generated at the end of each period.

Period (n)	Principal at Start of the Period	Interest Amount i_A	End-of-Month balance	Interest Payment	Principal Payment	Total Payment
1	$24,000	$720	$24,720	$720	$4,000	$4,720
2						
3						
4						
5						
Total						

12. Use the information provided in the previous problem, now consider that Eddie will be making 6 payments, at an interest rate of 3% per month and that he will only pay 50% of the interest amount generated at the end of each period. Use compound interest.

Period (n)	Principal at Start of the Period	Interest Amount (iA)	End-of-Month balance	Interest Payment	Principal Payment	Total Payment
1	$24,000	$720	$24,720	$360	$4,000	$4,360
2						
3						
4						
5						
6						
7						
Total						

Future Value Problems

13. A freshman student in your discipline is planning to continue his studies to grad school. Considering that the pay salary will increase according to an inflation rate of 4% per year determine how much he could be paid 10 years after graduation if he is offered a salary of $92,000 per year.

14. Use a 5% interest rate compounded each year to perform the following calculations. John's father gave him $22,000 on his 10th birthday. When he turned 11, he received $2,000; when he turned 12, he received $3,500; when he turned 13, he received $2,000; when he turned 14, he received $9,500; and when he turned 15, he received $12,000. Calculate the future value for John's 20th birthday.

15. Mary has a small cottage upstate, and she often rents it when she is not on vacation. Her yearly maintenance records indicate that she spent $5,500 in 2016, $6,500 in 2017, and $7,500 in 2018. If the year is 2024, calculate the total maintenance cost for the given flow in today's money. Consider a yearly interest rate of 8%.

16. Joe plans to retire in 25 years. If he wants to have $757,000 in his savings account when he retires, determine how much he needs to invest today if the savings pays 5% per year.

17. Determine the future value of tuition in 10 years assuming a present value of $50,000 and an interest rate of 5% per semester.

18. As a gift, Matthew's father invested $50,000 into a savings account on Matthew's fifth birthday to save for his college education. The savings account pays a yearly nominal interest rate of 7% compounded monthly. Today is Matthew's 15th birthday and his father wants to know what the current worth of his investment.

Present Value Problems

19. Determine the value of $1,000,000 today in 1900's money; consider an interest rate of 3% per year.

20. A company bought a piece of land for a new development in 2010. They believed that they would be able to develop a new apartment complex in 2020. However, due to the financial environment of 2020 they had to sell the land, for a price of $5,500,000, to cover some of their debt. Determine their purchase cost in 2010 using a yearly interest rate of 10%.

21. After saving for several years a man was able to save $2,500,000. Determine the number of periods he saved if the interest rate was 7.5% per year and the initial investment was $125,000.

22. I want to retire in 25 years with a savings value of $800,000. What should be the interest rate if I'm able to invest $100,000 today?

Nominal and Effective Interest Rates Problems

23. Determine the interest rate for weekly, monthly, and quarterly periods using a 20% yearly interest rate.

24. Mike can either select a payment scheme that charges a monthly 2% interest, or a bi-weekly 0.9% interest rate per period. Select the best choice by calculating the effective interest rate for both alternatives. (Hint: use 26 bi-weekly periods in a year)

25. Determine the yearly effective interest rate for a credit card with a 0.06% interest rate per-day.

26. Determine the continuous interest rate per year for the credit card with a 0.06% interest rate per-day.

27. I am buying a used car from my neighbor. The car is in great condition, and he is willing to sell it to me for $13,000. Unfortunately, I do not have the money, but I do want to buy the car. Therefore, I went to a bank and asked for a loan. The bank gave me the offer that I could pay them back the $14,500 loan at the end of six months.

    a. Determine how much interest I would be paying.

    b. Determine the nominal interest rate per year that I would be charged.

    c. Determine the effective interest rate per year that I would be charged.

    d. If my friend John gives me an effective interest rate of 6% per year, should I select this option instead of the offer from the bank?

28. Maggie decided to acquire a payday loan. The amount she borrowed was $350, and she agreed to pay back $425 the next payday (Maggie receives 12 payments in one year).

    a. Calculate what nominal interest rate per year (i_n) the lender is charging.

    b. Calculate what effective interest rate per year (i_{eff}) the lender is charging.

    c. If Maggie maintained this loan ($350) for a full year considering the effective interest rate determine how much would Maggie owe at the end of the year, assume she was unable to make any payments.

Interest Amount and Interest Rate

29. Determine the interest amount and the interest rate from borrowing $200 and from paying $250 after one month.
30. Brianna is buying a used car from Chloe. Chloe's car is in great condition, and she is willing to sell her car to Brianna for $17,000 (similar cars are worth $20,000).

    Therefore, Brianna went to the local bank to ask for a loan. The bank gave her the option of borrowing the $17,000 today and paying one lump sum of $22,000 in three years.

    a. Determine the interest amount for the loan.

    b. Determine the yearly interest rate for the loan.

    c. Brianna's uncle could lend her the money, but he would charge her a 5% interest rate per year. Should Brianna borrow from her uncle?

References

[1] 'Money makes money. And the money that makes money makes more money.'" QuoteFancy.https://quotefancy.com/quote/773277/Benjamin-Franklin-Money-makes-money-And-the-money-that-makes-money-makes-more-money (accessed Jan. 11, 2021).

[2] "History BankFive." https://bankfive.com/Our-Difference/Who-We-Are/History (accessed Jan. 12, 2021).

[3] "$0.05 in 1850 → 2020 CPI Inflation Calculator." https://www.officialdata.org/us/inflation/1850?amount=0.05 (accessed Jan. 12, 2021).

[4] "See the Average College Tuition in 2020-2021," *US News & World Report*. https://www.usnews.com/education/best-colleges/paying-for-college/articles/paying-for-college-infographic (accessed Jan. 12, 2021).

[5] "$0.05 in 2015 → 2020 CPI Inflation Calculator." https://www.in2013dollars.com/us/inflation/2015?amount=0.05 (accessed Jan. 12, 2021).

Chapter 3 Uniform and Non uniform (Gradient) Series Flows

Paula's Retirement Plan Story

IMG 3.1

Like many recent graduates, Paula was eager to finally enjoy the fruits of her time at college. She just received her degree and had a good job offer for $70,000.

Her college years had been difficult in many aspects, and having financial stress did not make it any easier.

Paula was ready to start her professional life but had to contend with her newly acquired financial responsibilities. She knew she had to plan for paying back her student loans and new SUV. In addition, she wanted to enjoy herself by doing some leisure activities.

On her first day of work, she met her boss and talked about her training; then she met with human resources to get information regarding payroll and benefits. One of the many surprises came when Paula asked about her net income after taxes and other deductions, and discovered it was much lower than expected: only $2,112 every two weeks. Therefore, her $70,000 salary after taxes and deductions had become $50,688. In other words, she would not receive about 28% of the income described in her job offer.

To make matters worse, the HR representative asked Paula how much she wanted to contribute to her 401(k) program. At this time Paula was thinking, I just lost 28% of my income and now you want to take more money out of my paycheck? Are you serious? I have several expenses already; I need to pay my student loans, my new car, car insurance, gas, and other living expenses. Besides, I also wanted to travel for the holidays. How can I afford an extra expense?

The HR representative explained that the 401(k) was a retirement plan where her employer contributes to her retirement, and she could also make voluntary contributions to increase her retirement funds. She described that one of the advantages of having a 401(k) was that this deduction was not taxable at the time of the investment but was taxable at the time of her retirement when she withdrew money. Moreover, her 401(k) was invested in the stock market, and it could increase its value over time.

Paula considered that she was too young at this time to think about retirement, so she decided to make the small contribution of $70 every paycheck.

Reading Activity: Responding to Paula's Retirement Plan Story
What is a retirement fund?
When should Paula start saving for her retirement?
What is the effect of having a series of investments and compound interest?
Was Paula's decision investment, correct? Should she have invested a higher or a lower amount?

Learning Outcomes

After covering this chapter, the student should be able to do the following:

1. Learn the relevance of uniform series of payments, also known as annuities.
2. Learn how to find the equivalence factors for annuities to a present and to a future value.
3. Learn how to calculate annuities for present values for infinite periods.
4. Lean how to calculate a cash flow series with increasing and decreasing flows, also known as gradients, by a fixed amount or by a fixed percentage.
5. Learn how to calculate present value and future value from uniform and nonuniform cash flow series.

Using Engineering Economy Factors

When solving complex engineering economy problems, one may have to perform a series of calculations. In such a case, it may be difficult to follow a series of formulas because a complex problem can require a sequence of multiple formulas, which would require memorization of each formula to find a solution.

Therefore, instead of expressing a problem with a series of engineering economy formulas it is more convenient to express such problems using equations, known as factors. These factors express the information described in the different formulas. To clarify this point, consider a future value from a present formula in Equation 2.6. Here, the different formula components are the variable we want to determine Future F given the known variables Present P, interest rate i, and n periods. This could be read as determining a Future given a Present.

Equation 2.1: Future for period n from a present value

$$F = P(1 + i)^n$$

The factor for Equation 2.6 is expressed as:

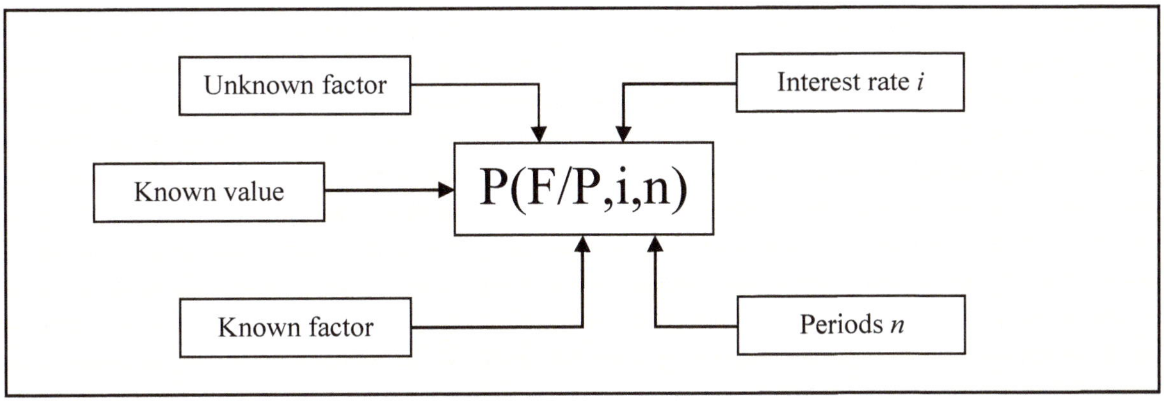

IMG 3.2

Hence, the factor for future value given a present value of \$100 for 30 periods at a 6% interest rate per period is represented as F = \$100(F/P, 6%, 30). Similarly, calculating a present value from a future value of \$100 for 15 periods at a 3% interest rate per period is represented as P = \$100(P/F, 3%, 15).

Introduction: Uniform Flow Series, Annuities

Uniform flow series, also known as annuities, represent a constant and continuous flow of investments or withdrawals made for a determined or indefinite duration. Finite duration is when the number of *n* periods is known such as in the case of buying a smart phone, while indefinite duration occurs frequently for infrastructure projects such as an airport or a bridge that may not have a known duration. Furthermore, some retirement funds such as Social Security provide an income for the retiree's life span.

A Word of Caution

The concepts described in this chapter -annuities and gradients- are some of the most relevant calculations in an engineering economy. The correct understanding of these concepts is very important as they are the foundation for the concepts described in the following chapters. Therefore, it is key that students generate a clear understanding of these topics.

Example 3.1 describes a situation where a person makes a series of payments for a loan.

Example 3.1

Michael bought a guitar for \$1,200. The guitar store offer was the following: no money down and make six monthly fixed payments of \$225.00. Fig. 3.1 shows the cash flow drawing for this transaction.

Example 3.1 (Continued)

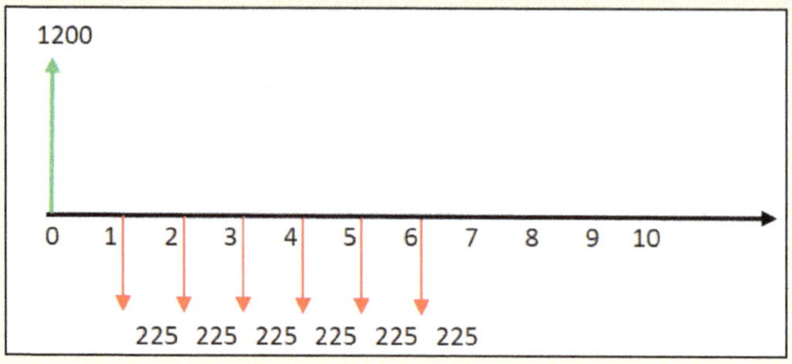

Fig. 3.1. Cash flow Example 3.1.

From Example 3.1 it is important to observe the following:

1. *The money borrowed is positioned at time zero.* This is because at time zero, the borrower receives the credit, and this is when the cash flow starts.
2. Most importantly, *annuities start one period after the present value, in other words, one period after the money was borrowed.*
3. The money borrowed is represented in the cash flow with an upward arrow because the borrower received the credit for buying the guitar.
4. Guitar payments are represented with a downward arrow as they are negative for the borrower.
5. All payments are for a fixed amount and are performed continuously for the *n* periods.

Annuities may also be calculated for future values. For example, consider a situation where a person invests in a retirement fund every month for a series of years. Here, the investor may be interested in learning how much how much would be accumulated from the investment and the accrued interest, as shown in Example 3.2.

Example 3.2

17 years ago, John started saving $1,000 every year for seventeen consecutive years in a college's fund to help pay for his son's college education. Develop the cash flow for his savings.

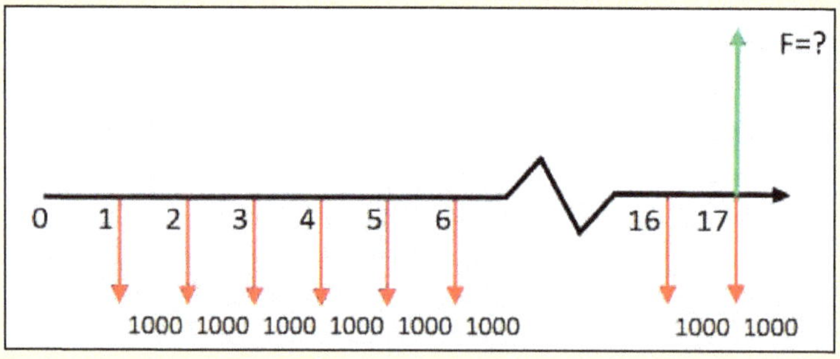

Fig 3.2. Example 3.2 College fund investment

For Example 3.2 we need to observe the following:

1. The money invested is represented with a downward arrow as it is negative for the investor.
2. Most importantly, *the last investment occurs in the same period as the future value occurs.*
3. The money saved in the future is represented with an upward arrow as it is positive for the investor.
4. Annuities have a fixed amount.

Calculation of Future Value from a Uniform Series of Payments

Consider the cash flow described in Example 3.2 and the future from a present equation 2.6.

Equation 2.6 Future from a present

$$F = P(1+i)^n$$

Deposit of $1,000 from year 0 and taking it to year 5, we obtain the following equation:

Equation 3.1 Future from a present for 5 periods

$$F = P(1+i)^5 = \$1000\,(1+i)^5$$

Note that the exponent value is 5 because there are five periods from year 0 to year 5.

To generate a formula for a series of equal deposits the variable P is replaced by the letter A to represent an annuity. To clarify this description, consider a situation where an investor makes five equal deposits starting in year 1 and ending in year 5, as seen in Fig. 3.3.

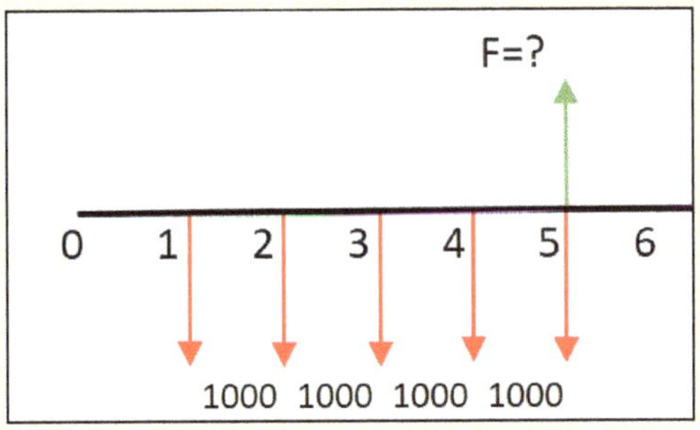

Fig. 3.3 Series of five equal payments.

Equation 3.2 Future from an annuity series from period 1 to period 5

$$F = A(1+i)^{n-1} + A(1+i)^{n-2} + A(1+i)^{n-3} + A(1+i)^{n-4} + A$$

Note that the last annuity A in year 5 is not multiplied by $(1 + i)$ since it occurs at the same time as the future value. When multiplying both equation sides by $(1 + i)$ it results in the following

Equation 3.3: Future from an annuity series multiplied by (1+i)

$$(1+i)F = A(1+i)^n + A(1+i)^{n-1} + \cdots + A(1+i)^{2-1} + A(1+i)$$

After factoring A and subtracting Equation 3.2 from Equation 3.3 it results in the following:

Equation 3.4: Future value from an annuity

$$F = \frac{A[(1+i)^n - 1]}{i}$$

The factor for Equation 3.4 is read as a future from an annuity, given an interest rate *i* for *n* periods.

Example 3.3

Consider the information provided in Example 3.2. John saved $1000 per year for seventeen consecutive years in a college fund to help pay for his son's college education. Determine how much he has in his account consider that the fund pays a yearly interest rate of 5%.

Using Equation 3.4

$$F = \frac{A[(1+i)^n - 1]}{i}$$

$$F = \frac{\$1000[(1+0.05)^{17} - 1]}{0.05} = \frac{\$1000 * 1.2920}{0.05} = \frac{\$1{,}292}{0.05} = \$25{,}840.36$$

Therefore, from Example 3.3 we observe that the future value from the series of payments is $25,840.36, an amount that is higher than the amount obtained from multiplying the annuities amount by the number of periods: $1,000 * 17 = $17,000. This demonstrates the effect of compound interest.

Exercise

Determine the future value after investing $1,000 for 10 years with a yearly interest rate of 4%.

Answer: $12,006.11

From Equation 3.4 we can also determine the formula for an annuity value from a future value.

Equation 3.5: Annuity from a future value

$$A = \frac{F * i}{[(1+i)^n - 1]}$$

Equation 3.5, (A/F, i%, n) which reads as annuity from a future from an annuity is given an interest rate i for n periods.

Example 3.4

Mary wants to have $1,000,000 in her 401(k) for her retirement. Her 401(k) account pays 10% per year and she wants to retire in 30 years. How much should she save each year?

$$A = \frac{F * i}{[(1 + i)^n - 1]}$$

$$A = \frac{1{,}000{,}000 * 0.1}{[(1 + 0.1)^{30} - 1]} = \frac{100{,}000}{16.4494} = \$6{,}079.24$$

Example 3.4 also shows that the future value of $1,000,000 is a higher amount than the amount obtained from multiplying the- annuity value times the number of periods $\$6{,}079.24 * 30 = \$182{,}377.20$, showing the effect of compounding interest rates.

Exercise

Determine how much Dan's dad should save every year for Dan's college education if he plans to save every year for 15 years and he wants his savings to reach $240,000. Consider an interest rate of 4% per year.

Answer $ $11,985.86

Calculation of Present Value from a Uniform Series of Payments

Considering Equations 2.6 and 3.4

$$F = P(1 + i)^n$$

$$F = \frac{A[(1 + i)^n - 1]}{i}$$

After substituting F from Equation 2.6 in Equation 3.4, we obtain the following results:

$$P(1+i)^n = \frac{A[(1+i)^n - 1]}{i}$$

Therefore, the annuity value, A, from a present value, P, is calculated as

Equation 3.6: Annuity from a present value

$$A = \frac{P(1+i)^n * i}{[(1+i)^n - 1]}$$

Equation 3.6 uses the following notation, (A/P. *i*%, n), which reads as an annuity from a present value, given an interest rate *i* for n periods.

Example 3.5

Josh bought a used car for $15,000. Determine how much he must pay each month for 24 months. The car dealer will charge a yearly interest rate of 6% per year.

We need to consider the monthly interest rate by diving the yearly interest rate by the number of months in a year.

$$i = \frac{6\%\ per\ year}{12\ months} = 0.5\%$$

Equation 3.6 to calculate the annuity value.

$$A = \frac{P(1+i)^n * i}{[(1+i)^n - 1]}$$

$$A = \frac{\$15{,}000(1+0.005)^{24} * 0.005}{[(1+0.005)^{24} - 1]} = \frac{\$15{,}000 * 0.0056}{0.1271} = \$15{,}000 * \ 0.0443$$

$$A = \$664.71$$

It is important to note how a compound interest rate affects the calculation of an annuity from a present value. Contrary to the results from Examples 3.3 and 3.4, the present value from Example 3.5, $15,000, is lower than the value of multiplying the value of the annuities by the number of periods, $28,684.32.

Exercise

Kishore wants to know how much he would have to pay each month for 60 months for a new machine for his business. The machine cost of $55,000 and the seller will charge a 4% interest rate per year.

Answer: $1,012.91

Similarly, using Equation 3.6 the present value, P, from the annuity value, A, is calculated as

Equation 3.7: Present value from an annuity

$$P = \frac{A * [(1+i)^n - 1]}{(1+i)^n * i}$$

Equation 3.7 uses the notation, (P/A. i%, n), which is read as present value from an annuity, given an interest rate i for n periods.

Example 3.6

Luke spends about $40 a month on a sports streaming service. He is thinking about cancelling this service and putting this money in a savings account that pays 12% per year. Determine the present value of this investment if he invests this amount for a period of a year and a half.

The expense per month is $40.

The monthly interest rate is calculated as 12% per year / 12 months.

Therefore, the monthly interest rate = 1%.

The number of periods is n = 18 months

$$P = \frac{A * [(1+i)^n - 1]}{(1+i)^n * i}$$

$$P = \frac{\$40 * [(1+0.01)^{18} - 1]}{(1+0.01)^{18} * i} = \$40 * \ 16.4002 = \$656$$

As described in Example 3.5, Example 3.6 also shows that the present value, $6,560.08, is lower than the value obtained from multiplying the value of the annuities by the number of periods, $7,200.

Exercise

Determine the present value and future value of paying streaming services for ten years, with a monthly cost of $30. Consider an interest rate of 6% per year.

Answer: PV= $2,702.20

FV= $4,916.38

Using Excel to Solve Examples PV, FV, and Annuities

In this section we will solve the previous examples using Excel. It should be noted that Excel assumes you entered a balanced equation. To clarify this point let us consider the following example. Assume you invested \$100 at time 0 and want to calculate its future value after three years. Excel will require you to enter the present value in year 0, the interest rate per period, and the number of periods.

Assume that you entered an investment amount using a positive sign; Excel will then display the result of the future value with a negative sign. This is because Excel assumes that when you entered a positive value you receive a positive flow of money, and in turn it considers that the future value represents a payment, and therefore a negative flow of money. On the other hand, if the invested amount is entered with a negative sign, it will calculate the future amount using a positive sign.

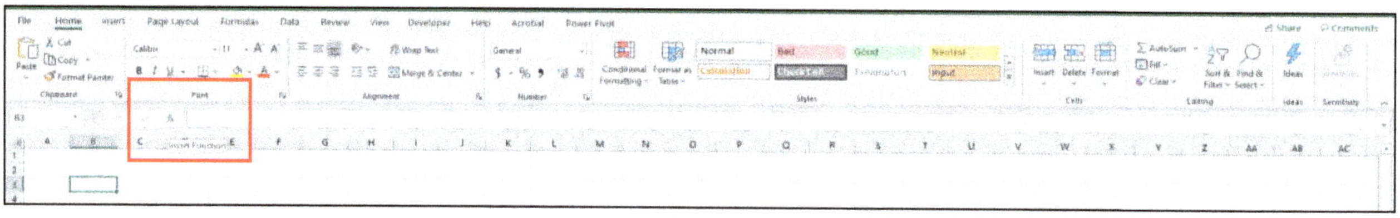

Fig. 3.4. Excel function f_x icon.

The Excel formula to find a future is FV. It can be called by typing "=FV" in any cell or by finding the Future function from clicking the f_x icon formula, as shown in Fig. 3.4. This formula is used in Example 3.3 Excel solution.

The Excel function to find an annuity is PMT, which stands for payment. It can be called by typing "=PMT" in any cell or by finding the PMT function from the f_x icon function. This function is used in Example 3.4 Excel solution.

Example 3.3: Excel solution

John saved \$1000 per year for 17 consecutive years in a college fund to help pay for his son's college education. He wants to know how much he has in his account today. Consider a yearly interest rate of 5%.

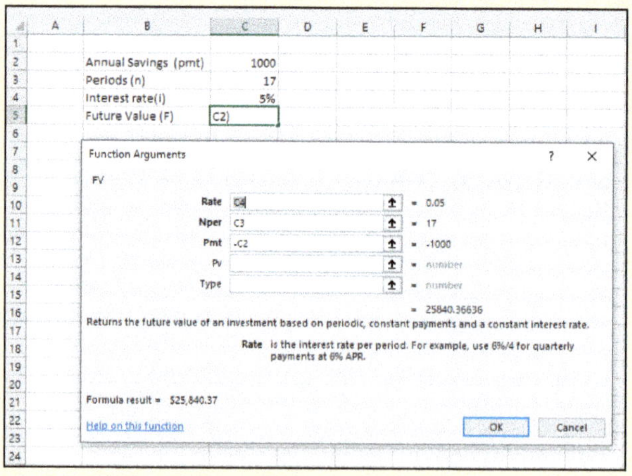

Fig. 3.5. Excel solution Example 3.3

Example 3.4: Excel solution

Mary wants to have $1,000,000 in her 401(k) for her retirement. Her savings account pays 10% per year, and she wants to retire in 30 years. How much should she save each year?

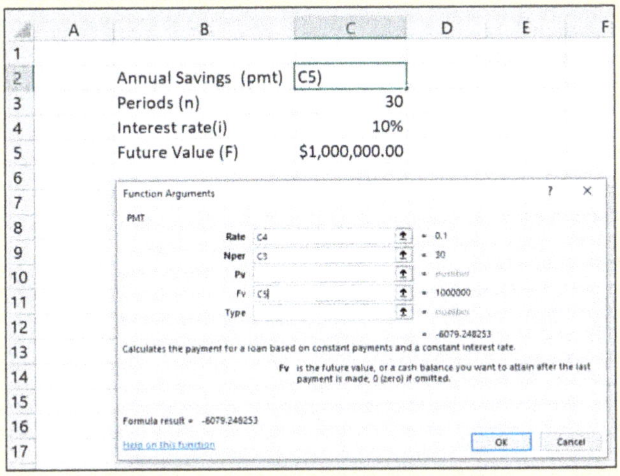

Fig. 3.6. Example 3.4 Excel solution.

The MS Excel solution for Example 3.5 uses the FV function described in the Excel solution for Example 3.4. The only difference is that we now enter the present value PV value instead of the future value FV.

Example 3.5: Excel solution

Josh bought a used car for $15,000. Determine how much he has to pay each month for 24 months. The car dealer will charge a yearly interest rate of 6% per year.

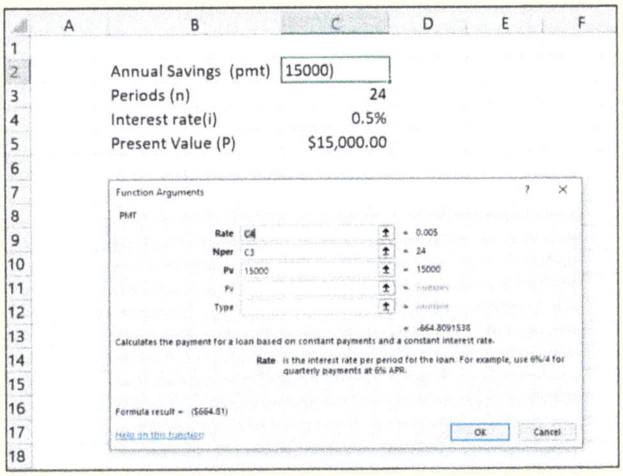

Fig. 3.7. Example 3.5 Excel solution.

The Excel solution for Example 3.6 uses the PV function to calculate the present value.

Example 3.6: Excel solution

Luke spends about $40 a month on a sports streaming service. He is thinking about cancelling this service and putting this money in a savings account that pays 12% per year. Determine the present value of this investment if he invests this amount for a period of a year and a half.

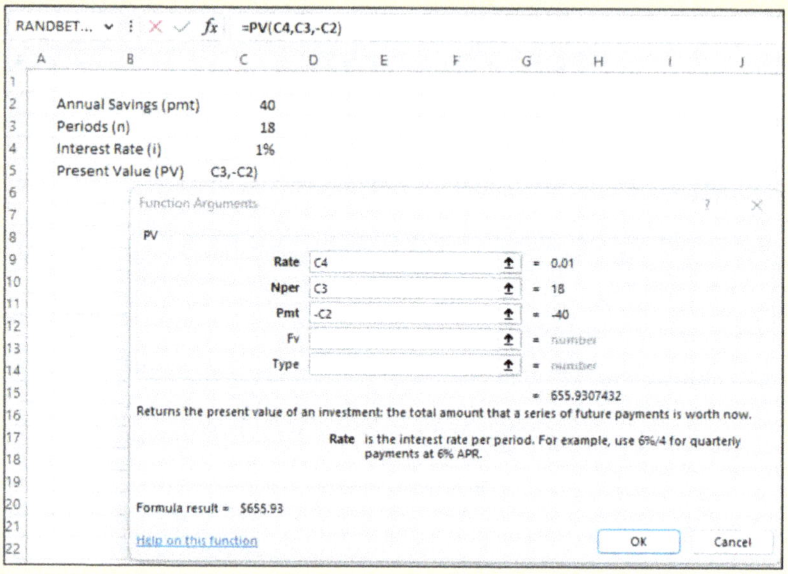

Fig. 3.8. Present value from an annuity Example 3.6

Mid-Chapter Key Points Review

1. Future values occur *in the same time period* as the last annuity.
2. Future values have *a higher value* than the value obtained from multiplying annuities by the number of periods.
3. Present values *start one period before* the annuities; for example, if you borrow today, you would make your first payment in the next period.
4. Present values have *a lower* amount than the value obtained from multiplying annuities by the number of periods.
5. Excel can be used to calculate equivalence values between present, future, and annuities.

Annuity Calculation for Infinite Flows

Some projects do not have a determined number of periods; in fact, there are some projects where the goal of the project is to provide funds for an infinite number of periods. Some of these could be government infrastructure projects such as bridges or dams, or funds for charitable purposes such as scholarship funds or research funds.

To develop this formula, consider Equation 3.6 when n → ∞

Equation 3.6: Annuity from a present value

$$A = \frac{P(1+i)^n * i}{[(1+i)^n - 1]}$$

Taking Equation 3.6 and dividing the numerator and denominator by $(1+i)^n$ to maintain balance and taking the limit when n goes to infinite we obtain.

Equation 3.8: Annuity from a present value and dividing by $(1+i)^n$

$$\lim_{n\to\infty} A = P * \frac{\frac{(1+i)^n * i}{(1+i)^n}}{\frac{[(1+i)^n - 1]}{(1+i)^n}}$$

Next, the equation is simplified and evaluated for a limit taken to an infinite value:

Equation 3.9: Equation 3.8 to a limit to infinite.

$$\lim_{n\to\infty} A = P\frac{\frac{i}{1}}{1 - \frac{1}{(1+i)^n}} = Pi$$

Therefore, the annuity from a present value for an infinite number of periods is determined as

Equation 3.10: Annuities from a present value for an infinite number of periods

$$A = Pi$$

To clarify this equation, consider a situation in which an investor only withdraws the interest amount generated from his investment without withdrawing money from the principal. Therefore, the principal amount remains constant and in turn the interest amount generated also remains constant.

The following example compares annuity values derived from having a set number of periods and an infinite number of periods.

Example 3.7

XYZ State University created a fund to provide scholarships to students with disabilities. The fund has $5,700,000. XYZ State wants to provide scholarships every semester. The financial planner for XYZ State will invest the money in a fund that pays 5% per year.

Example 3.7

Since the funds will be provided every semester the interest rate is 2.5% per semester.

a. Determine how much XYZ State University would have available each semester considering a time horizon of 30 years (60 semesters).

Annuity = $5,700,000 (A/P, 2.5%, 60) = $5,700,000* 0.0324 = $184,680 each semester

b. Determine how much XYZ State University would have available each semester considering an infinite time horizon.

$$A = Pi$$

Annuity = $5,700,000 * 2.5% = $5,700,000 * 0.025 = $142,500 each semester

From Example 3.7 we learned:

1. As the number of periods increases the annuity values decrease. Therefore, annuities for perpetual projects have a lower value.
2. Annuities for perpetual projects can be calculated using Equation 3.10.

Calculation of Annuities with Inconsistent Values and Irregular Time Periods

There are situations where annuities have inconsistent values and irregular time periods following. For example, an investor may decide to increase a savings amount for a single period or to defer a payment. Examples 3.8 and 3.9 consider such situations.

Example 3.8

Michael invested $5,000 in his savings account for seven consecutive years; after making these investments Michael left his investment in his savings account for three years (without adding or withdrawing any money). In year 10 Michael wants to use this money to fix his house, determine how much he has. Consider a yearly interest rate of 5%.

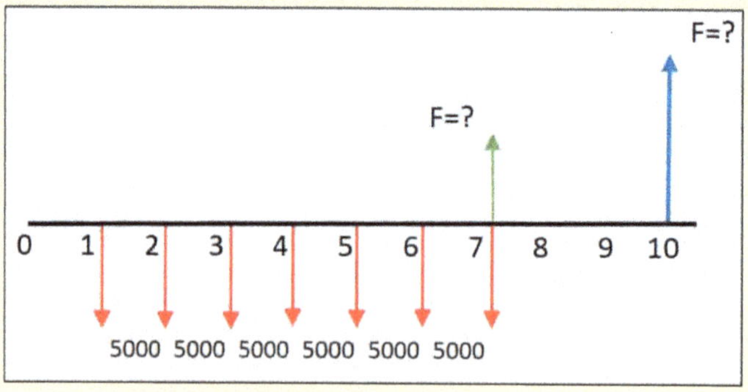

Figure 3.9. Cash flow Example 3.8.

Example 3.8

Annuity = $5,000 for years 1 to 7

$i = 5\%$

First, let us calculate the amount invested during the first seven years.

Future year 7 = $5,000(F/A, 5%, 7) = $5,000 * (8.142) = $40,710.04

Next, the future value for year 10 is calculated by taking the present value of year 7 to a future value in year 10.

Future Year 10 = $40,710.04 * (F/P, 5%, 3) = $40,710.04 * (1.158) = $47,142.18

The Excel solution for Example 3.8 is shown.

Example 3.8 Excel Solution

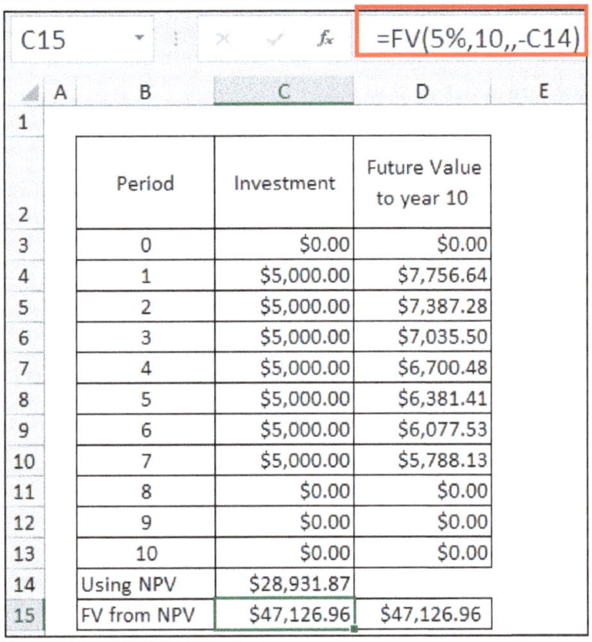
C15 =FV(5%,10,,-C14)

Period	Investment	Future Value to year 10
0	$0.00	$0.00
1	$5,000.00	$7,756.64
2	$5,000.00	$7,387.28
3	$5,000.00	$7,035.50
4	$5,000.00	$6,700.48
5	$5,000.00	$6,381.41
6	$5,000.00	$6,077.53
7	$5,000.00	$5,788.13
8	$0.00	$0.00
9	$0.00	$0.00
10	$0.00	$0.00
Using NPV	$28,931.87	
FV from NPV	$47,126.96	$47,126.96

Fig. 3.10. From top to bottom, NPV function (calculates the PV from year 0 for the seven investments FV function (brings the PV year 0 to year 10).

The NPV function in cell C14 adds the present in time 0, cell C3 with the Net Present Value from the series of payments starting in year 1 "=NPV(5%,C4:C10)+C3." Therefore, the PV for year 0 must be added to the value obtained from the NPV function.

The NPV calculates the present value, the future value is calculated by bringing the present value from year 7 to a future value in year 10.

Example 3.9

Maddy made seven deposits in total, six for a value of \$5,000 and one for a value of \$15,000 Fig 3.11. The length of each period is six months. Maddy would like to make one withdrawal at the end of the year 6; determine how much she will be able to withdraw. Consider a yearly interest rate of 6%.

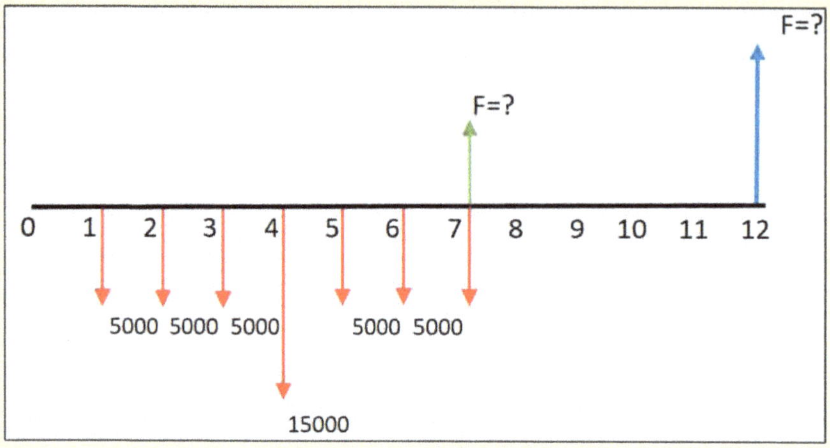

Figure 3.11. Cash flow Example 3.9.

1. First, recalculate the interest rate per period.

   Since the yearly interest rate is 6% and the length of each period is six months, and since there are two periods of six months in a year, the interest rate per six-month period is $i = {}^{6\%}/_{2} = 3\%$

2. Second, calculate the future value for period 7.

   Future period_7 = \$5,000(F/A,3%,7) = \$5,000 * (7.662) = \$ 38,310

3. Third, calculate the future value from the seven deposits from period 7 to period 12 by calculating a future from a present.

   Future period_{12} = \$ 38,310 * (F/P, 3%, 5) = \$ 38,310 * (1.159) = \$ 44,401.29

4. Fourth, take the remaining \$10,000 from period 4 to period 12.

   Future period_{12} = \$10,000 (F/P,3%,8) = \$10,000 * (1.267) = \$ 12,670

5. Fifth, add both values, which are now in year 12.

   Future period_{12} = \$ 44,401.29 +\$ 12,670 = \$ 57,071.29

The Excel solution for Example 3.9 is shown.

Example 3.9 Excel Solution

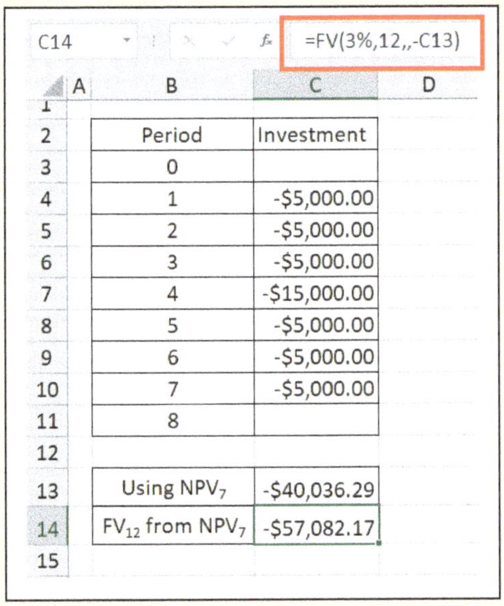

C14 =FV(3%,12,,-C13)

	Period	Investment
2	Period	Investment
3	0	
4	1	-$5,000.00
5	2	-$5,000.00
6	3	-$5,000.00
7	4	-$15,000.00
8	5	-$5,000.00
9	6	-$5,000.00
10	7	-$5,000.00
11	8	
12		
13	Using NPV_7	-$40,036.29
14	FV_{12} from NPV_7	-$57,082.17
15		

Figure 3.12: Cash flow Example 3.9

As in Example 3.8, we used the NPV function to determine the net present value for years 1 to 7.

The calculated PV is set for period 0, one period before the first annuity in period 1, using the following function in C13: =NPV(3%,C4:C10)

Then the function to calculate the future value for year 6 (period 12) is described in cell C14.

Exercise

Solve Example 3.9 considering that Maddy would like to make a withdrawal of $15,000 in period 12. Determine how much can she withdraw in period 16.

Answer $47,363.85

Non-uniform Series Flow (Gradients)

There are several everyday situations that require non-uniform calculations. For example, due to inflation some products often increase by a percentage each year. Similarly, certain products may increase or decrease by a certain amount for every period; for instance, consider a situation where a monthly rent price is increased each year by $100 dollars.

Cash flows that change (either increase or decrease) **by a certain amount** per period are called **arithmetic gradients** denoted by **"G"**; on the other hand cash flows that change (either increase or decrease) **by a percent rate** are called **geometric gradients** denoted by **"g."** Consider an application of this concept in Example 3.10.

Example 3.10

After working for a full year Mike saved $1000 in his savings account. He continued saving at the end of each year for the next five years by increasing his savings by $100, as shown in Fig. 3.13 and the following table. Draw the cash flow for his savings.

Year	0	1	2	3	4	5
Amount	$0	$ 1,000	$ 1,100	$ 1,200	$ 1,300	$ 1,400

Example 3.10 (continued)

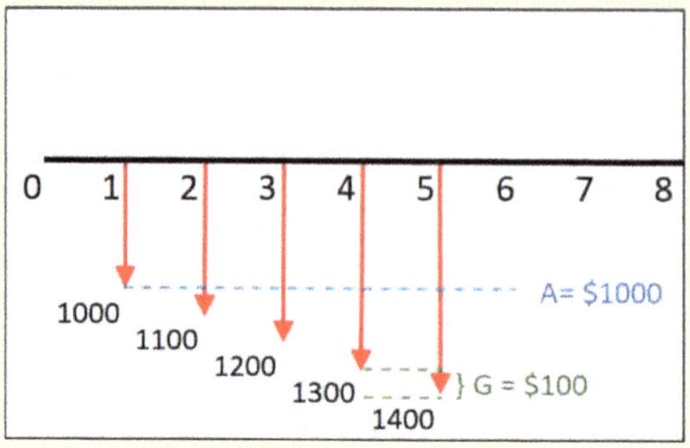

Figure 3.13: Example 3.10

From Example 3.10 we learned the following:

1. The value of the gradient G is the amount difference between the different periods. Therefore, the gradient for period 1 is 0G, for period 2 it is 1G, for period 3 it is 2G, and for period n it is (n-1)*G.
2. The gradient occurs for the same number of periods as the annuity. In the example there are five annuities and five gradients. However, the increased value due to the gradient is first shown in the second period.
3. As in the case of annuities the gradient starts at the first period.
4. The present value at time 0 for Example 3.10 is calculated by adding the present value from the annuity plus the present value from the gradient.

Calculation of a Present Value from an Arithmetic Gradient

As described, the gradient G is constant; the gradient for period 1 is 0G, for period 2 it is 1G, for period 3 it is 2G, and for period n it is (n-1)*G. Therefore, calculating the present value in time 0 for the gradient values from Example 3.10, we have the following:

Equation 3.11: Present value (PV) for the gradient in Example 3.10

$$P = G\left(\frac{1}{(1+i)^2}\right) + 2G\left(\frac{1}{(1+i)^3}\right) + \cdots + (n-1)G\left(\frac{1}{(1+i)^n}\right)$$

Factoring G from Equation 3.11 and multiplying both sides of the equation by $(1+i)^n$

Equation 3.12: Factoring G from Equation 3.11 and multiplying both sides of the equation by (1+i)[1]

$$iP = G\left[\frac{1}{(1+i)^1} + \frac{1}{(1+i)^2} + \cdots + \frac{1}{(1+i)^{n-1}} + \frac{1}{(1+i)^n}\right] - G\left[\frac{1}{(1+i)^n}\right]$$

After subtracting Equation 3.11 from Equation 3.12 and then simplifying, we obtain the following equation to calculate a present value from an arithmetic gradient:

Equation 3.13: Present from an arithmetic gradient

$$P = G\left[\frac{(1+i)^n - in - 1}{i^2(1+i)^n}\right]$$

Therefore, solving an increasing arithmetic gradient requires that we add the Present value from the annuity plus the present value from the gradient.

*Equation 3.14: Present value for an **increasing** flow calculated by adding the present value from an annuity and the present value from the arithmetic gradient.*

$$P = \frac{A * [(1+i)^n - 1]}{(1+i)^n * i} + G\left[\frac{(1+i)^n - in - 1}{i^2(1+i)^n}\right]$$

It should be noted that the first part of Equation 3.14 is only present from an annuity P/A in Equation 3.7, and the second part refers to a present value given a gradient P/G in Equation 3.13. Moreover, for increasing gradients P/A it is important to note that the value of the annuity corresponds to the constant annuity value for all cash flows. Example 3.11 shows an increasing gradient while Example 3.12 shows a decreasing gradient.

Example 3.11: Solution from Example 3.10

After working for a full year Mike saved $1000 in his savings account, he continued saving at the end of each year for the next five years. Each year he increased the savings amount by $100. Use a 3% interest rate per year.

Using equation 3.14

P=P/A+P/G

P = $1,000 (P/A, 3%, 5) +$100(P/G,3%,5)

$1,000 (P/A, 3%, 5) Equation 3.9

$$P = \frac{A * [(1+i)^n - 1]}{(1+i)^n * i} = \frac{\$1000 * [(1.03)^5 - 1]}{(1.03)^5 * 0.03} = \$1000 * 4.580 = \$4{,}580$$

$100 (P/G,3%,5) Equation 3.15

$$P = G\left[\frac{(1+i)^n - in - 1}{i^2(1+i)^n}\right] = \left[\frac{(1.03)^5 - (0.03 * 5) - 1}{(0.03)^2 * (1.03)^5}\right] = \$100 * 8.889 = \$888.9$$

P=P/A+P/G

P=$4,580+ $888.9 = $5,468.9

Example 3.11: Excel Solution

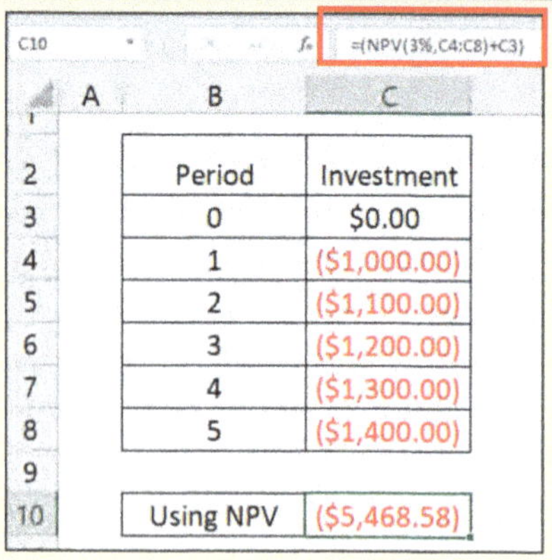
C10 =(NPV(3%,C4:C8)+C3)

Period	Investment
0	$0.00
1	($1,000.00)
2	($1,100.00)
3	($1,200.00)
4	($1,300.00)
5	($1,400.00)
Using NPV	($5,468.58)

Fig. 3.14. Example 3.11

As in Examples 3.7 and 3.8, the NPV function calculates the net present value for years 1 to 5.

Since the PV for value for period 0 is zero the PV value for period 0 does not require any extra calculations.

This same approach may be used for solving Example 3.12.

In contrast to Example 3.11, where the arithmetic gradient increases, there are situations where an arithmetic gradient decreases. In this case the present value is calculated by subtracting the present value, calculated from the gradient Equation 3.13, from the present value from the annuity Equation 3.7. This calculation is shown in Equation 3.15. It is important to note that for decreasing gradients the value of the annuity corresponds to the highest annuity value.

*Equation 3.15: Present value for a **decreasing** flow calculated by subtracting the arithmetic gradient from the PV from an annuity.*

$$P = \frac{A * [(1+i)^n - 1]}{(1+i)^n * i} - G\left[\frac{(1+i)^n - in - 1}{i^2(1+i)^n}\right]$$

This situation is shown in Example 3.12.

Example 3.12

Michael, a small hotel owner, invested in a new solar roof to reduce his electric bill. Since solar panels are expensive, he spread his investment out over 10 years. His electric bill for year 1 was $2,500 and was reduced by $250 each year. Determine the present value of his electric bill for year 0. Consider an interest rate of 4% per year.

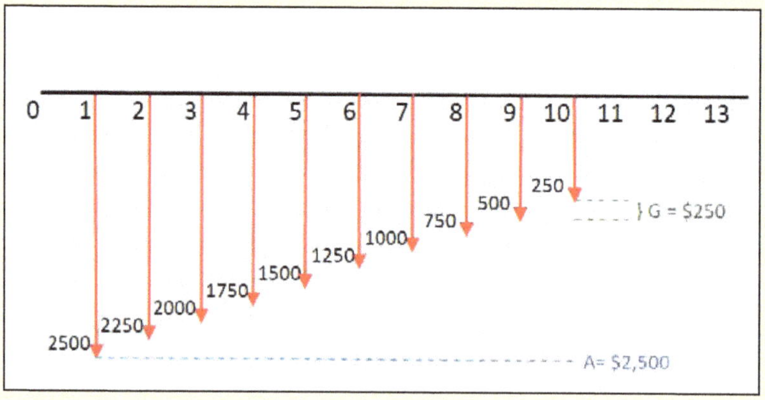

Fig. 3.15. Cash flow Example 3.12

Example 3.12 (continued)

$$P_0 = PA - PG$$

P_0 = \$2,500 (P/A, 4%, 10) - \$250(P/G,4%,10)

Using Equation 3.15,

\$2,500 (P/A, 4%, 10)

$$P = \frac{A * [(1+i)^n - 1]}{(1+i)^n * i} = \frac{\$2{,}500 * [(1.04)^{10} - 1]}{(1.04)^{10} * 0.04} = \$2{,}500 \; * \; 8.1108 \; = \; \$20{,}277.23$$

\$250(P/G,4%,10)

$$P = G\left[\frac{(1+i)^n - in - 1}{i^2(1+i)^n}\right] = \$250 * \left[\frac{(1.04)^{10} - (0.04 * 10) - 1}{(0.04)^2 * (1.04)^{10}}\right]$$

$$P_0 = \$250 \; * 33.8813 \; = \; \$8{,}470.33$$

$$P_0 = \$20{,}277.23 - \$8{,}470.33 = \$11{,}806.90$$

From Examples 3.11 and 3.12 we learned the following:

1. For increasing (growing) gradients, the annuity is taken from the first annuity for all cash flows plus the gradient value.
2. For decreasing (shrinking) gradients, the annuity is taken from the first annuity value and the gradient amount is then subtracted. The highest value of the annuity is often the first value in the gradient.

Calculation of an Annuity from an Arithmetic Gradient

Equating Equations 3.7 (present from an annuity) and 3.13 (present from an arithmetic gradient) it is possible to derive the formula to calculate an annuity from a gradient by simply dividing both equations by $(1+i)^n * i$:

$$\frac{A * [(1+i)^n - 1]}{(1+i)^n * i} = G\left[\frac{(1+i)^n - in - 1}{i^2(1+i)^n}\right]$$

Equation 3.16: Annuity from a gradient

$$A = G\left[\frac{(1+i)^n - in - 1}{[i(1+i)^n - i]}\right]$$

Let us recalculate Example 3.10. Assume that this time Mike wants to find an equivalent annual investment for the cash flow.

Example 3.13

From the information in Example 3.11. After working for a full year Mike saved $1000 in his savings account. He continued saving at the end of each year for the next five years and increased his savings amount by $100 each year. Use a 3% interest rate per year.

First, let us consider the annual value from the annuity as described in the problem statement annuity = $1,000. Second, determine the Annuity from the gradient by using Equation 3.16.

$$A=\$1000 + \$100\ (A/G,3\%,5)$$

$$A = G\left[\frac{(1+i)^n - in - 1}{[i(1+i)^n - i]}\right] = \$100\left[\frac{(1+0.03)^5 - 0.03 * 5 - 1}{[0.03(1+0.03)^5 - 0.03]}\right] = \$100 * 1.9409$$

$$A = \$194.09$$

Therefore, the annuity is calculated by adding both annuity values,

$A = \$1,000 + \$194.09 = \$1,194.09$

In Example 3.11 the calculated present value is P = $5,468.9.

Let us confirm our calculation by calculating the present value from the annuity value from the recent calculation.

$$P= A(P/A,3\%,5)$$

$P= \$1,194.09 * 4.580 = \$5,468.93$

Calculation of a Present Value from a Geometric Gradient

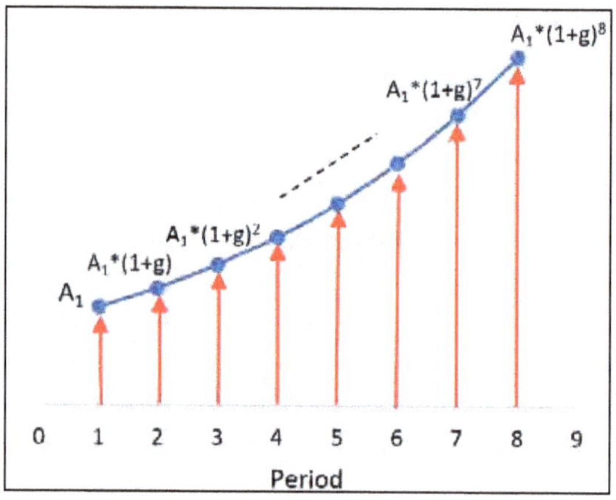

Fig. 3.16. Geometric gradient for 8 periods.

Geometric gradients increase or decrease by a percentage for every period. Figure 3.16 shows a geometric gradient. For geometric gradients, the notation uses a lowercase letter **g** to describe the rate of change between periods. In addition, we use the variable A_1 to describe the initial annuity value.

To derive the P from a geometric gradient **g** formula, consider the following geometric gradient for eight periods. Here, the initial annuity value is A_1 and is multiplied by an amount $1 + g$ for the following periods.

To determine the present value for **g,** consider the present from a future formula. In this case the present value from the initial value A_1 is calculated as:

$$P = A_1\left(\frac{1}{(1+i)^1}\right)$$

Therefore, let us calculate the series of present values from the future by adding all future values:

Equation 3.167: Present from a series of future values

$$P = A_1\left(\frac{1}{(1+i)^1}\right) + A_1\left(\frac{1(1+g)^1}{(1+i)^2}\right) + A_1\left(\frac{1*(1+g)^2}{(1+i)^3}\right) + \cdots + A_1\left(\frac{1*(1+g)^{n-1}}{(1+i)^n}\right)$$

Then, by multiplying both sides of the equation by $\frac{1+g}{1+i}$ and subtracting this result from Equation 3.16, the resulting equation calculates the Present value from a geometric gradient, Equation 3.17.

Equation 3.178 Present from a geometric gradient for i ≠ g

$$P = A_1\left[\frac{1-(1+g)^n(1+i)^{-n}}{i-g}\right] \; for\; i \neq g$$

This factor is described for an equation as (P/A,g,i,n), where *i* is the interest rate and g is the geometric gradient rate.

On the other hand, there are situations where the interest rate and the gradient have the same value. In such instances Equation 3.17 may not be used and the present value is obtained by multiplying the annuity value by the number of periods and by dividing by $1 + i$, as observed in Equation 3.18.

Equation 3.19: Present from a geometric value for i = g

$$P = \left[\frac{A_1 * n}{(1+i)}\right] \; for\; i = g$$

A calculation of the geometric gradient is described in Example 3.14.

Example 3.14

George, a young child, started selling lemonade during the summer and earned \$100 in his first year. He continued selling lemonade for the following 8 summers. Each summer, his earnings increased by 20%. Determine the present value of his earnings in year 0, considering a yearly interest rate of 2%.

Example 3.14 (continued)

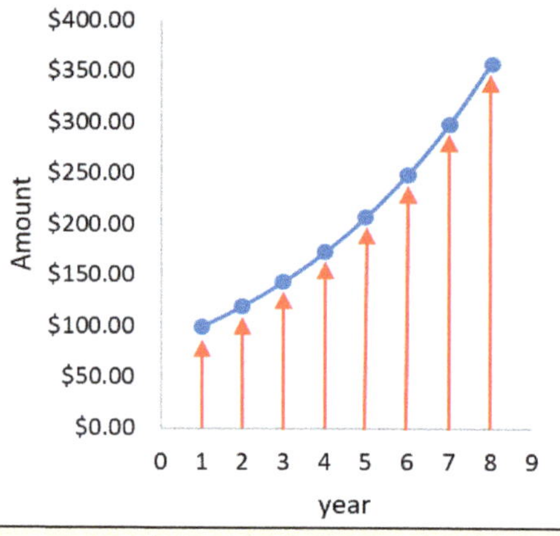

year	value
0	
1	$100.00
2	$120.00
3	$144.00
4	$172.80
5	$207.36
6	$248.83
7	$298.60
8	$358.32

Fig. 3.17. Example 3.14 geometric gradient.

i = 2% and g = 20% considering Equation 3.18

$$P = A_1\left[\frac{1-(1+g)^n(1+i)^{-n}}{i-g}\right]$$

$$P = \$100\left[\frac{1-(1+0.2)^8(1+0.02)^{-8}}{0.02-0.2}\right] \qquad P = \$100\left[\frac{1-4.29*0.85}{-0.18}\right]$$

$$P = \$100\left[\frac{-2.66}{-0.18}\right] = \$100*14.83 = \$1{,}483.25$$

From Examples 3.14 we learned the following:

1. For geometric gradients it is required to verify the gradient and interest rate values to select the correct equation.
2. In contrast with the arithmetic gradient calculation, only one factor is required to determine the present value for the gradient series.
3. The initial annuity value A_1 is required for calculating the gradient value.
4. Since Equations 3.17 and 3.18 calculate the present value, if a problem requires the calculation of annuity value or future value, one must first calculate the present value of the gradient and then calculate the annuity or future value from the present.
5. The MS Excel solution for Examples 3.10 to 3.14 use the MS Excel function NPV, as portrayed in the MS Excel solution of Example 3.9

What Did We Learn in This Chapter?

- How to develop the formulas to calculate annuities for present and future values and how to use these formulas
- How to calculate uniform and non-uniform flows
- How to calculate annuities for an infinite number of periods
- What gradients are and how they can be calculated?
- How to determine the number of periods for annuities and gradients
- How to use Excel to calculate annuities and cash flows

Review Questions

1. Why is it relevant to learn annuities?
2. What is the difference between annuities for a future and annuities for a present?
3. How many different gradients are there? What are their differences?
4. What is the difference in calculating an increasing and a decreasing gradient?
5. If you have an apartment lease, would you rather have the rent increase using a geometric or an arithmetic gradient?
6. When do annuities end when calculating future values from an annuity?
7. When do present values start when calculating a present from an annuity?
8. Do present values have *a lower* amount than the value obtained from multiplying annuities by the number of periods? If so, why or why not?
9. What is the difference between an annuity and a gradient?
10. Is there difference in the number of periods in annuities and gradients?

Chapter 3 Formulas

Equation 2.18: Future for period n from a present value

$$F = P(1+i)^n$$

Equation 3.19: Future value from an annuity

$$F = \frac{A[(1+i)^n - 1]}{i}$$

Equation 3.20: Annuity from a future value

$$A = \frac{F * i}{[(1+i)^n - 1]}$$

Equation 3.21: Annuity from a present value

$$A = \frac{P(1+i)^n * i}{[(1+i)^n - 1]}$$

Equation 3.22: Present value from an annuity

$$P = \frac{A * [(1+i)^n - 1]}{(1+i)^n * i}$$

Equation 3.23: Annuities from a present value for an infinite number of periods

$$A = Pi$$

*Equation 3.24: Present value for an **increasing** flow calculated by adding the present value from an annuity and the present value from the arithmetic gradient.*

$$P = \frac{A * [(1+i)^n - 1]}{(1+i)^n * i} + G\left[\frac{(1+i)^n - in - 1}{i^2(1+i)^n}\right]$$

*Equation 3.25: Present value for a **decreasing** flow calculated by subtracting arithmetic gradient from the PV from an annuity.*

$$P = \frac{A * [(1+i)^n - 1]}{(1+i)^n * i} - G\left[\frac{(1+i)^n - in - 1}{i^2(1+i)^n}\right]$$

Equation 3.18: Annuity from a gradient

$$A = G\left[\frac{(1+i)^n - in - 1}{[i(1+i)^n - i]}\right]$$

Equation 3.26 Present from a geometric gradient for i ≠ g

$$P = A_1\left[\frac{1 - (1+g)^n(1+i)^{-n}}{i - g}\right] \; for\; i \neq g$$

Equation 3.18: Present from a geometric value for i = g

$$P = \left[\frac{A_1 * n}{(1+i)}\right] \; for\; i = g$$

Chapter 3 Practice Problems

Use the following problems to practice your newly acquired knowledge. Answers to even problems are in the back of the book.

Annuities

1. Determine the future value of Paula's 401(k) savings described in the chapter's introduction. Assume Paula maintained her contribution of $70 for every paycheck. She is paid twice a month (assume 24 payments per year). Assume that she worked for 35 years and that her 401(k) paid 10% per year.

2. How much should Ben invest in an account today to make five yearly consecutive disbursements of $40,000 starting three years from today? Consider a yearly interest rate of 6%.

3. Determine the present value of investing 10 consecutive monthly investments of $100. Assume a yearly interest rate of 12%.

4. Joe plans to retire in 35 years. If he wants to have $7,570,000 in his savings account when he retires, determine how much he needs to invest in his monthly savings if the savings account pays 10% per year.

5. Perform a web search to find the current cost of a sports car. Determine the monthly cost for a period of 36 and 64 months, using a yearly rate of 3.25%.

6. John's father gave him $22,000 on his 10th birthday. When he turned 11, he received $2,000 and continued receiving $2,000 until the age of 16. Assume today is John's 20th birthday. Determine how much he has today, assuming a 5% interest rate per year.

7. Jonah made six consecutive yearly deposits of $10,000, starting in year 0, and he wants to make three equal withdrawals in years 8, 9, and 10. Determine the amount of these withdrawals. Assume an interest rate of 3% per year.

8. A savings account pays 10% per year. Jonah made five consecutive deposits of $550, starting in year 1. After completing these five deposits he made three deposits of $750, in years 8, 9, and 10, and did not withdraw or deposit any additional money. How much he will have in year 10?

9. Julia wants to determine how much she should invest in a fund every month to reach $550,000 for her retirement. She plans to start her investments one year from today and wants to retire in 25 years. Consider a yearly interest rate of 12%.

10. The Social Security Administration (SSA) takes a portion of a person's salary and adds it to an amount paid by the person's employer, with the purpose of providing monthly retirement benefits. Consider an interest rate of 2.5% per year and determine the following:

    a. How much would the Social Security collect from a person earning $50,000 per year? Assume a working life of 35 years and that the amount contributed to the SSA by the employee and its employer is 12% of the person's income.

    b. If a retired individual received $8,000 per month for 15 years, determine if the amount paid by the SSA was greater than the amount provided by the employee's contribution. Use the same interest rate as before.

11. Louis just retired and wants to sell his $750,000 house in New York and move to a more affordable overseas destination. His brother wants to buy his house and offers to pay him in 60 months with a 0.5% interest rate per month. Determine the monthly payment amount?

12. Jimmy bought a car for $37,000 with 84-month financing and a yearly interest rate of 5%. Determine how much he has to pay every month.

13. A state government wants to build a dam. Assume that the state government is considering a yearly interest rate of 20% and the dam has a cost of $160,000,000 with yearly operation costs of $1,500,000.

Gradients

14. A company needs $25,000 next year to pay their property taxes. Assume that property taxes in their county increase by 3% per year. Determine how much the company needs to deposit today in a trust fund to cover this expense for the next 10 years. Consider (a) an interest rate of 3% and (b) an interest rate of 6%.

15. Jeffrey's, a local convenience store, wants to know the future value of their marketing campaign in year 5. Jeffrey's spent $100,000 in year 1, $110,000 in year 2, $120,000 in year 3, $130,000 in year 4, and $140,000 in year 5.

16. Lenny wants to know how much he needs to have saved 15 years from today to cover for his son's tuition. His son is 2 years old today, and Lenny just made his first contribution to a savings fund that pays 3% per year. His first contribution was $5,000, and he plans to increase his savings contribution by either $500 or by 7% each year. Determine the future value of his savings using both contribution plans and describe which he should consider for maximizing his savings.

17. A small company owner wants to install solar panels to reduce his electric bill. Since solar panels are expensive, he plans on adding new panels every year. The expected electric bill for the next 10 years is shown. Determine the present amount needed to cover these savings; assume an interest rate of 2.5% per year. Note: Negative cost means that the electric company will buy the extra power.

Year	0	1	2	3	4	5	6	7	8	9	10
Cost	$2,000	$1,750	$1,500	$1,250	$1,000	$750	$500	$250	$0	-$250	-$500

18. John inherited $150,000 and maintained this amount for 25 years, in an account with an interest rate of 7% per year. He wants to withdraw X dollars next month and wants to increase the withdrawal amount by X every year. Determine the value of X assume; a time horizon of 10 years.

Chapter 4 Present Value Method

Paula's Rent versus Buy Story

IMG 4.1

Paula's first job had a short commute of 45 minutes, so she decided that there was no need to move due to the convenience, which thus saved her money.

After a few months on the job Paula's boss was pleased with her performance and mentioned this to her own boss, the VP of manufacturing. He told Paula's boss that they had a sudden need to fill a quality management position in their Midwest facility. After a short conversation between Paula's boss and the VP of manufacturing, they both decided that Paula was an excellent candidate for the job. Even though Paula's boss was sad to see her go, she thought that Paula and the company would benefit from this change. Paula was offered the quality management position and within a few days she left her job in the Northeast and moved to the Midwest.

One of the advantages of having a new job was that Paula's salary increased by $15,000 per year and her living costs decreased as the Midwest was more affordable. A house in her new location was about 60% of the price of a house back home, so she was excited about the idea of buying her own place.

However, Paula still had her student loans, car loan, and other expenses to contend with, and she did not want to commit to a new expense without having more certainty about her new job. She wondered, what would happen if after buying the house she did not like her new job? Or what if she would get relocated again? Or what if any other circumstance changed?

Therefore, Paula wanted to perform an analysis to know when it would be a good idea to stop paying for rent and start buying a house.

Reading Activity: Responding to Paula's Rent-versus-Buy Story
Should Paula compare buying versus rent? How could Paula make this comparison? What information is required to know the full cost of owning a house? What should be the time horizon for the analysis? Should her analysis be based on today's circumstances?

Learning Outcomes

After covering this chapter, the student should be able to do the following:

1. Learn how to set and perform present worth calculations from uniform and nonuniform annual flows
2. Learn how to select the best alternative among mutually exclusive alternatives using present worth
3. Learn how to perform present worth calculations for mutually exclusive alternatives with different lives

Introduction: Comparison Methods

In previous chapters we learned how to calculate equivalent amounts of money for different time periods and cash flows. For instance, we learned how to calculate a future value from a present value and how to calculate a future value from a series of payments. However, as described in this chapter's reading, there are situations in which we need to know the equivalent value of a cash flow, so we may evaluate and select among different alternatives.

Some of the most common methods used to compare alternatives are as follows:

- Present worth method
- Equivalent uniform annual flows
- Internal rate of return

This chapter considers the calculation of the present worth method; the other two methods are presented in the following chapters. It is important to note that if correctly performed, these comparison methods would select the same alternative. Furthermore, since these alternatives are mutually exclusive, only one option can be selected.

The present worth method is one of the more widely use methods for project evaluation. The logic of this method is to determine the equivalence at time 0 of the different cash flows from the future and then compare this equivalence to the initial investment. When the present worth is positive, it means that the project has a positive flow of money and therefore may be recommended. If the present worth is negative, it means that the project will lose money, and it may not be recommended. However, when this analysis is about cost we may select the project that offers the lowest cost. Furthermore, as previously described, the concept of the time value of money does not allow us to add/subtract cash flow values from different time periods since these amounts are affected by an interest rate.

Present Worth Calculation

The method of present worth or present value considers the comparison of two or more alternatives in today's value. For example, assume that today you want to compare the value of two cars, one with a cost of $22,000 and the other with a cost of $33,000. Since both amounts occur at the same time (today) it is easy to identify and select the alternative that has the lowest cost. However, this may not be as straightforward when we have a cash flow that incorporates additional financial information, such as operation costs, maintenance costs, fees, insurance, and salvage values. In addition, the term *present worth* is used to describe that the net flow considers positive and negative flows.

To clarify the car comparison example, perform a quick web research for the price of a regular gas vehicle and compare it to the price of an electric vehicle with similar class and size. As you may find in your search, the current sales price of the electric vehicle is higher than the cost of the gas vehicle.

However, electrical car companies claim that gas savings, maintenance, salvage value, depreciation, and tax incentives may offset the difference. Therefore, the present worth or present value calculation is performed by bringing all the elements of the cash flow to the present value in time 0. Let us consider the following examples to clarify this point.

Example 4.1

Michael just bought a car for $22,000. Michael estimates that the car operation and maintenance cost will be $250 per month. He believes that he will keep the car for six years and that after using the car he will be able to resell it for $4,000. Consider a 6% yearly interest rate. Draw the cash flow and determine the present worth for his car:

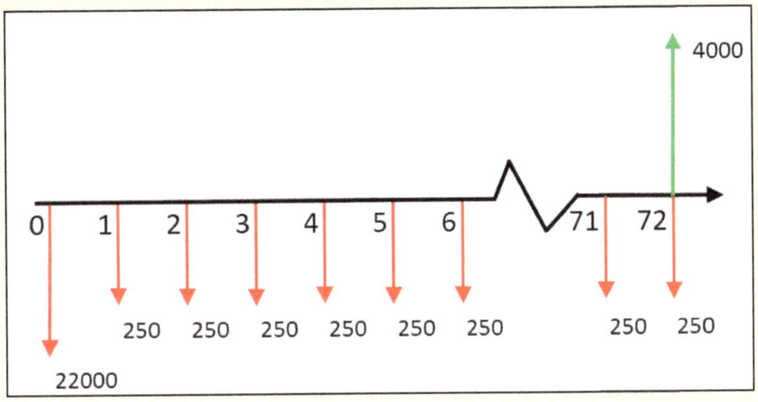

Fig. 4.1. Cash flow for buying a car and keeping it for 72 months.

Car cost = $22,000 (performed in time 0)

Monthly operation and maintenance = $250

Salvage value = $4,000

Interest rate = 6% per year = 6% / 12 = 0.5% per month

n = 6 years = 72 months

Present worth = -$22,000 – $250 (P/A, 0.5%,72) + $4,000 (P/F, 0.5%,72)

Present worth = -$22,000 – $250 (60.34) + $4,000 (0.6983)

Present worth = -$22,000 – $15,085 + $2,793.2 = -$34,291.80

Now let us consider the following example for an electric car.

Example 4.2

Tom, Michael's brother, bought an e-car for $33,000. He received an instant rebate of $2,000 from New York state. Unfortunately, he spent $2,500 setting up a charging station at home, but car operation and maintenance costs are estimated to be only $40 per month. Tom is also keeping the car for six years, and he thinks that he will be able to resell it for $9,000. Consider a 6% yearly interest rate.

Example 4.2 (Continued)

Draw the cash flow and determine the present worth for his car.

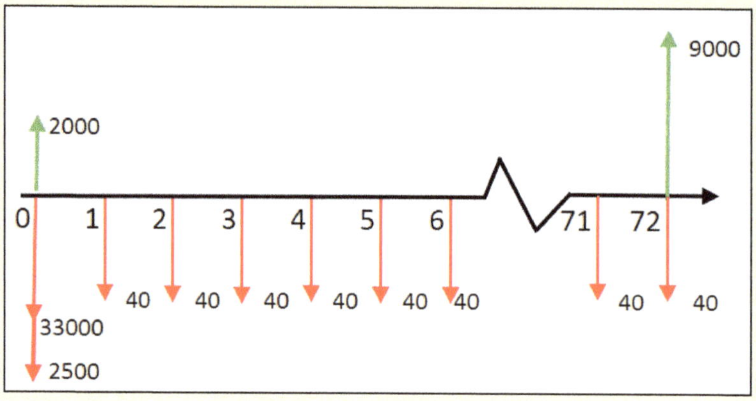

Fig. 4.2. Cash flow for an e-car.

Car cost = \$33,000 (performed in time 0)

Electric installation cost = \$2,500

Rebate from NY state = \$2,000

Net value in time 0 = -\$33,000 – \$2,500 +\$2,000 = -\$33,500

Monthly operation and maintenance = \$40

Salvage value = \$9,000

Interest rate = 6% per year = 6% / 12 = 0.5% per month

n = 6 years = 72 months

Present worth = -\$33,500 – \$40 (P/A, 0.5%,72) + \$9,000 (P/F, 0.5%,72)

Present worth = -\$33,500 – \$40 (60.34) + \$9,000 (0.6983)

Present worth = -\$29,628.90

From the information provided in these examples we can observe that e-vehicles are cheaper than gas vehicles. However, this textbook example might differ from a real-life scenario, as we would need to gather real costs and savings for the specific car options to verify this result the following.

In Examples 4.1 and 4.2 we learned:

1. Both flows are negative since cars are an expense. Unless you own a car-rental, taxi service or other car service cars generally do not generate an income.
2. Both flows are calculated to the present time in year 0.

3. Net flows for cash flows occurring at the same time can be obtained by adding values.
4. Cash flows from different time periods are brought to a present value in time 0 to generate a net present worth in time 0.
5. These two examples have the same number of time periods;, this is not always the case.

Finding Present Worth for Gradients

Consider the example described in the reading passage at the start of the chapter. Paula would like to know if she should rent or buy a house. In general, rent prices increase every year by either an amount or a percentage, and renting only gives the right to live for a time without acquiring ownership rights. Home buying, on the other hand, is more attractive as it provides ownership, which is one of the ways to build wealth. However, owning a house has additional costs such as maintenance and property taxes. Moreover, owning a house for a short period of time may be a bad financial decision as there are several expenses related to the process of buying a house such as lawyer fees, real state fees and mortgage costs.

Example 4.3

Paula will be renting a nice two-bedroom apartment for a monthly payment of \$1,500. Her lease stipulates that her rent will increase by 3% each year. Determine the PV for renting after keeping the apartment for 3 years, 10 years, and 20 years. Consider a 4% yearly interest rate.

$i = 4\%\ per\ year$

$g = 3\%\ per\ year$

A = \$1,500 per month or \$18,000 (It is solved per year as the compounding is per year.)

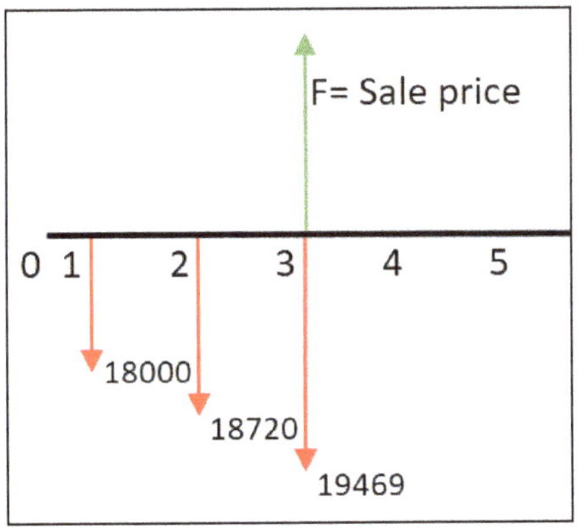

Fig. 4.3: Cash flow for buying a house for 3 years.

Consider the PV formula from a gradient described in chapter 3:

$$P = A_1\left[\frac{1-(1+g)^n(1+i)^{-n}}{i-g}\right]\ for\ i \neq g$$

Example 4.3 (Continued)

$i = 4\%\ per\ year$

$g = 3\%\ per\ year$

A = $1,500 per month or $18,000 (It is solved per year as the gradient is increased per year.)

Consider the PV formula from a gradient described in chapter 3:

$$P = A_1\left[\frac{1-(1+g)^n(1+i)^{-n}}{i-g}\right]\ for\ i \neq g$$

For n = 3

$$P = -\$18{,}000\left[\frac{1-(1+0.03)^3(1+0.04)^{-3}}{0.04-0.03}\right] = -18{,}000 * 2.8569 = -\$51{,}425.42$$

For n = 10

$$P = -\$18{,}000\left[\frac{1-(1+0.03)^{10}(1+0.04)^{-10}}{0.04-0.03}\right] = -18{,}000 * 9.2098 = -\$165{,}776.85$$

For n = 20

$$P = -\$18{,}000\left[\frac{1-(1+0.03)^{20}(1+0.04)^{-20}}{0.04-0.03}\right] = -18{,}000 * 17.5714 = -\$316{,}285.94$$

Consider that instead of renting Paula could buy a house with a $40,000 down payment and a $1,550 monthly payment, which includes property taxes, maintenance costs, and a mortgage payment.

Determine the PV after keeping the house for 3 years, 10 years, and 20 years. Assume that the net flow after reselling the house and paying what is left of the mortgage is $24,000 after 3 years, $114,000 after 10 years, and $339,000 after 20 years. Assume a 4% yearly interest rate.

The PV from buying, reselling the house, and the monthly mortgage payment brings all cash flow values to time zero. Fig. 4.3 shows the cash flow for 3 years. A similar flow with different salvage values would occur for 10 years and 20 years (Fig. 4.4).

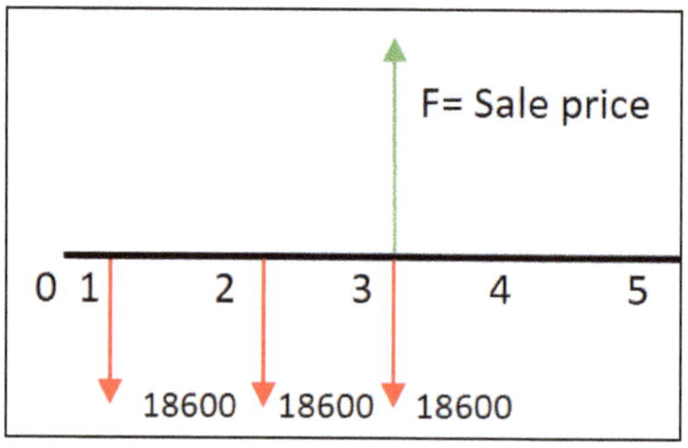

Fig. 4.4. Cash flow for buying a house 3 years

Example 4.3 (Continued)

Therefore, the problem data is the following:

$i = 4\%\ per\ year$

A = \$1,550 per month or \$18,600 (As before, we solve per year to match, previous calculation.)

Equation for n = 3 years

PW = -\$40,000 – \$18,600 (P/A, 4%, 3) + \$24,000 (P/F,4%, 3)

PW = -\$40,000 – (\$18,600 *2.775) + (\$24,000 * 0.8890) = -\$70,279.00

Equation for n = 10 years

PW = -\$40,000 – \$18,600 (P/A, 4%, 10) + \$114,000 (P/F, 4%, 10)

PW = -\$40,000 – (\$18,600 *8.111) + (\$114,000 *0.6756) = -\$113,846.20

Equation for n = 20 years

PW = -\$40,000 – \$18,600 (P/A, 4%,20) + \$339,000 (P/F, 4%,20)

PW = -\$40,000 – (\$18,600 *13.590) + (\$339,000 *0.4564) = -\$138,054.40

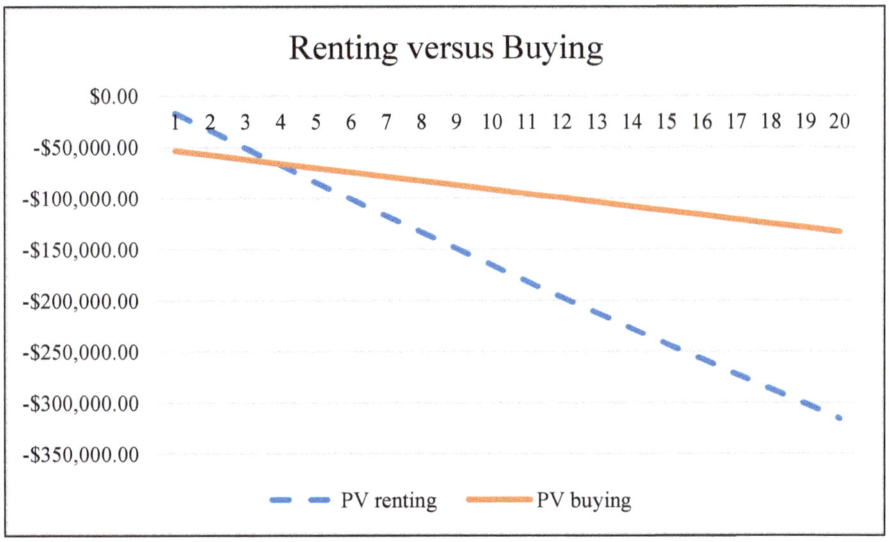

Fig. 4.5. Renting versus Buying.

As observed in Fig. 4.5. the PW from renting is higher for a short period of time than buying. This example presents a rough estimate for this problem, as owning a house has additional costs than the mortgage, such as property tax, scheduled maintenance, and emergency repairs. In addition, the sales price for the house at a given time depends on a variety of factors, including but not limited to supply and demand. However, it should be noted that owning a house is a great way to build wealth.

In Example 4.3 we learned the following:

1. The cost of renting was lower than owning a house and selling it after a short period of time. You need to consider that this only considers the information provided in the problem statement.
2. As before, all cash flows were calculated to the present time in year 0.
3. We considered a yearly analysis for the first part of the problem as the value of the rent increased every year; otherwise, we would have performed a monthly analysis.

Example 4.4 shows how to calculate present worth from a series of annuities and a decreasing gradient.

Example 4.4

Jim invested $5,000 in a small business and received $1,000 in years 1, 2, and 3. In year 4 there was a fire, so he did not receive money from his investment in years 4 and 5. The business reopened in year 6, and he received benefits for $2,000. In year 7 he received benefits for $1,500, in year 8 he received benefits for $1,000, in year 9 he received benefits for $500, and in year 10 he did not receive any benefits, so they decided to close the business. Determine the PV for his investment. Consider a yearly interest rate of 6%.

According, the description the cash flow is as follows:

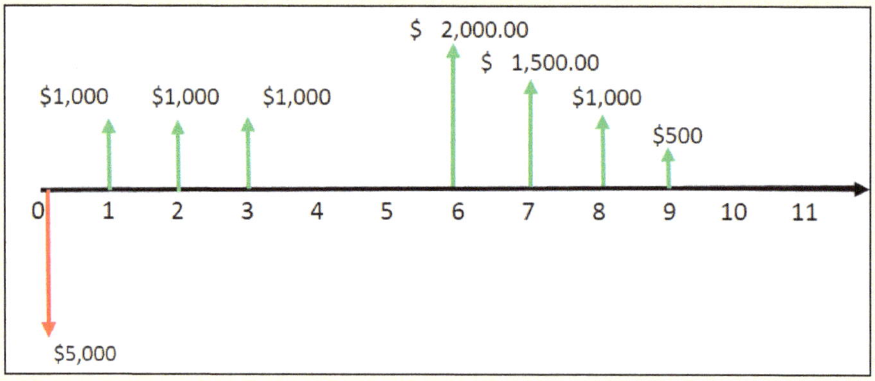

Fig. 4.6. Cash flows for Example 4.4.

PV = -$5,000 + $1000(P/A,6%,3) + [$2000(P/A,6%,4)*(P/F,6%,5)] – [$500(P/G,6%,4)*(P/F,6%,5)]

PV = -$5,000 +$1000(2.6730) + [$2000(3.4651)*(0.7473)] – [$500(4.9455)*(0.7473)]

PV = -$5,000 + $2,673.01 + $5,178.65 - $1,847.78 =$1,003.88

In Example 4.4 we learned the following:

1. When there are two flow series, it is required to perform each PW calculation separately by bringing one cash series flow at a time.
2. For the second flow, we should observe that the PV from the gradient is first calculated for year 5, one period before the start of the gradient, and then the PV from year 5 is brought to year 0 by using the factor present from a future

Exercise

Nicholas invested $116,500 in his bank account today (year 0). He is planning on withdrawing an amount (t) in year 5, (t+150) in year 6, (t+300) in year 7, and (t+450) in year 8. Use a yearly interest rate of 8% and determine the value of t.

Answer: t = $ 47,646.70

Different Lives Analysis

Previous Examples 4.1, 4.2, and 4.3 show two alternatives with the same duration. However, there are situations where alternatives have different lives, such as in cases when an option lasts longer than another. In these situations, it is required to find the lowest common multiple for both life durations and then perform the financial analysis using the lowest common multiple for both alternatives. For example, if one alternative has a life of three years and the other alternative has a life of four years, then both alternatives would be analyzed for a period of 12 years.

Example 4.5

Determine the present worth for the following two options; assume a yearly interest rate of 6%.

Option	Cost	Service Life in Years	Savings Per Year	Salvage Value
A	$4,484	3	$1,800	0
B	$7,985	6	$2,700	$1,500

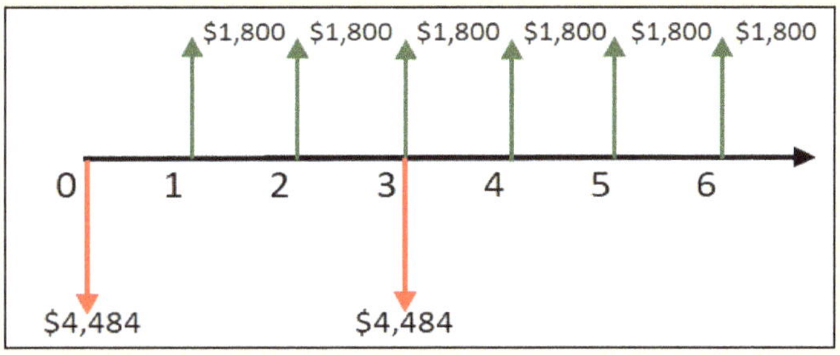

Fig. 4.7. Cash flow for Example 4.5 option A.

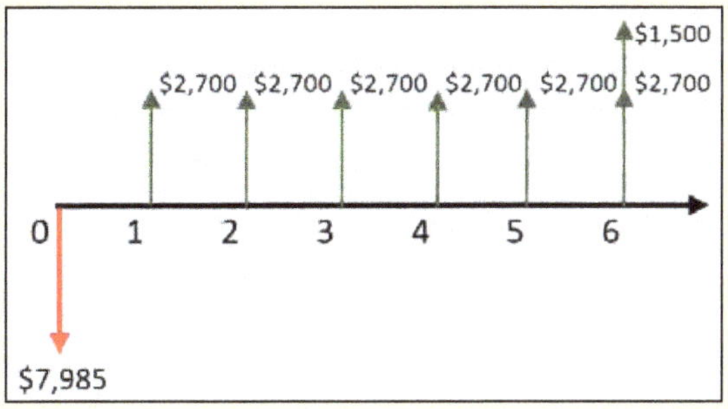

Fig. 4.8. Cash flow for Example 4.6 option B.

Example 4.5 (Continued)

PW option A: Since this option only lasts for three years it needs to be replaced with an identical replacement after three years.

$$PW_A = -\$4{,}484 + \$1{,}800\ (P/A,6\%,6) - \$4{,}484\ (P/F,6\%,3)$$

$$PW_A = -\$4{,}484 + \$1{,}800\ (4.9173) - \$4{,}484\ (0.8396) = \$602.33$$

PW option B:

$$PW_B = -\$7{,}985 + \$2{,}700\ (P/A,6\%,6) + \$1{,}500\ (P/F,6\%,6)$$

$$PW_B = -\$7{,}985 + \$2{,}700\ (4.9173) + \$1{,}500\ (0.7050) = \$6{,}349.22$$

In Example 4.5 we learned the following the following:

1. We considered the lowest common multiple for both life durations. In this case since 6 is a multiple of three we consider the value of 6.
2. For simplicity, it is assumed that the replacement cost is identical to the original cost and that it will have the same duration. In reality, this may not be entirely accurate due to market changes and technological advances.

Exercise

Recalculate Example 4.5 considering a life of two years for option A and a life of four years for option B.

Answers: Option A = -$2,237.55
Option B = $2,558.92

The previous analysis shows that the best alternative is option B. However, in a real-life project this selection may not be entirely correct because it is unlikely that option A will be replaced by an identical item with an identical cost in three years. Furthermore, due to technical changes in three years there might be additional options in the market to replace option A.

Therefore, to improve this analysis we could do the following:

- Forecast, with the highest level of accuracy possible, what will happen in the future and then estimate the different alternatives that might be available in market.
- Perform the analysis considering a set number of years or the lowest amount of time for the different lives considering what the salvage value for the option with the longest life horizon. This is shown in the Example 4.6.

Example 4.6

Determine the present worth for the following two options, use a yearly interest rate of 6%.

Option	Cost	Service Life in Years	Savings Per Year	Salvage Value After 3 Years
A	$4,484	3	$1,800	0
B	$7,985	6	$2,700	$2,000

Solve this problem considering a life of 3 years.

PW Option A. Since this option only lasts for 3 years it does not need to be replaced.

PW Option A

$$PW_A = -\$4,484 + \$1,800\ (P/A,6\%,3)$$

$$PW_A = -\$4,484 + \$1,800\ (2.6730) = -\$4,484 + \$4,811.42 = \$327.42$$

PW Option B

$$PW_B = -\$7,985 + \$2,700\ (P/A,6\%,3) + \$2,000\ (P/F,6\%,3)$$

$$PW_B = -\$7,985 + \$2,700\ (2.6730) + \$2,000\ (0.8396)$$

$$PW_B = -\$7,985 + \$7,217.1 + \$1,679.2 = \$911.3$$

In Example, 4.6, option B is still a better option since the PW value is higher than option A.

In Examples 4.5 and 4.6 we learned the following:

1. When items have different life durations we can consider the lowest common multiple for both life durations. Assuming an identical replacement cost for the option which will be replaced sooner.
2. We can also adjust the duration of the project to the lowest life or to a different life value. However, performing a specific duration may require acquiring additional information for replacement costs and salvage values (if they exist).

Indefinite Number of Periods

As previously described in Chapter 3, there are circumstances where a project has an infinite number of periods. Consider the following equation from Chapter 3:

Equation 4.1 (Equation 3.12 from chapter 3): Annuities from a present for an infinite number of periods

$$A = Pi$$

From here we can rewrite the formula as:

Equation 4.2: Present from annuities for an infinite number of periods

$$P = \frac{A}{i}$$

The following example compares annuity values derived from having a set number of periods and an infinite number of periods.

Example 4.7

A local university invested in a new recreational center. The university estimates that they will spend $485,000 in maintenance costs per year. The university would like to invest $3,500,000 into a perpetual investment fund to pay for this maintenance expenditure. Assume that the amount they will be investing will be monthly compounding. Determine the interest rate that the fund should pay.

First, consider the equation to calculate a present value from annuities for an infinite number of periods:

$$P = \frac{A}{i} \text{ which turns to } i = \frac{A}{P}$$

Therefore,

$$i = \frac{A}{P} = \frac{485{,}000}{3{,}500{,}000} = 13.85\%$$

From Chapter 2 we divide the yearly interest rate by the number of monthly periods per year:

$$i_n = 13.85\% \,/12 = 1.1541\% \; per\ month$$

Effect of Interest Rates in Present Worth

In Chapter 2 we described how *economical equivalence* is a term used to describe how two different amounts of money are equivalent even when the amounts differ. Moreover, we learned that interest rates make these amounts equivalent, and the selection of a proper interest rate has relevant consequences to the present worth analysis in the selection of mutually exclusive alternatives. Let us consider the following alternatives.

Example 4.8

Determine the present worth for the following two options. Consider using a yearly interest rate of 5% and a yearly interest rate of 10%.

Option	Cost	Year 1	Year 2	Year 3	Year 4	Year 5	Year 6
A	-$1,900	1400	250	250	250	250	250
B	-$1,800	300	300	300	300	300	1500

Example 4.8 (Continued)

Using a 5% interest rate we obtain the following:

PW option A

$PW_A = -\$1,900 + \$250\ (P/A,5\%,6) + \$1,150\ (P/F,5\%,1)$

$PW_A = -\$1,900 + \$250\ (5.076) + \$1,150\ (0.9524)$

$PW_A = -\$1,900 + \$1,269 + \$\ 1,095.26 = \$464.26$

Using a 5% we obtain the following:

PW option B

$PW_B = -\$1,800 + \$300\ (P/A,5\%,6) + \$1,200\ (P/F,5\%,6)$

$PW_B = -\$1,800 + \$300\ (5.076) + \$1,200\ (0.7462)$

$PW_B = -\$1,800 + \$1,522.8 + \$895.44 = \$618.24$

Using a 10% we obtain the following:

PW option A,

$PW_A = -\$1,900 + \$250\ (P/A,10\%,6) + \$1,150\ (P/F,10\%,1)$

$PW_A = -\$1,900 + \$250\ (4.355) + \$1,150\ (0.9091)$

$PW_A = -\$1,900 + \$\ 1,088 + \$1045.46 = \$234.21$

PW option B

$PW_B = -\$1,800 + \$300\ (P/A,10\%,6) + \$1,200\ (P/F,10\%,6)$

$PW_B = -\$1,800 + \$300\ (4.355) + \$1,200\ (0.5645)$

$PW_B = -\$1,800 + \$1843.5 + \$677.4 = \$183.9$

Exercise

Mandy will retire next year and wants to know how much should be invested in an eternal fund to provide for her monthly expenses. Assume that her living cost are \$4,500 and consider that her retirement fund has a yearly interest rate of 8%.

Answer : \$675,000

As observed in Example 4.8, the value of the present worth for each option and the alternative selection is affected by the interest rate. Option B had a higher present worth value than option A when calculated using a 5% yearly interest rate. On the other hand, option B had a lower present worth value than option A when using a 10% yearly interest rate. This situation can be explained by of the following:

1. High interest rates tend to favor projects that have higher benefits during the early stages of the project.
2. On the other hand, low interest rates tend to favor projects with higher benefits during the later stages of the project.

This can be observed in Figure 4.9. The table describes the present worth values for options A and B based on the interest rates from 1% to 20%. The plot shows how PW changes for each option for the different interest rates. Furthermore, the table and the plot show that these PW values become negative at 13% for option B and at 17% for option A.

i	PWA	PWB
1%	$687	$1,069
2%	$628	$946
3%	$571	$830
4%	$516	$721
5%	$464	$618
6%	$414	$521
7%	$366	$430
8%	$321	$343
9%	$277	$261
10%	$234	$184
11%	$194	$111
12%	$155	$41
13%	$117	-$24
14%	$81	-$87
15%	$46	-$146
16%	$13	-$202
17%	-$20	-$255
18%	-$51	-$306
19%	-$81	-$354
20%	-$110	-$400

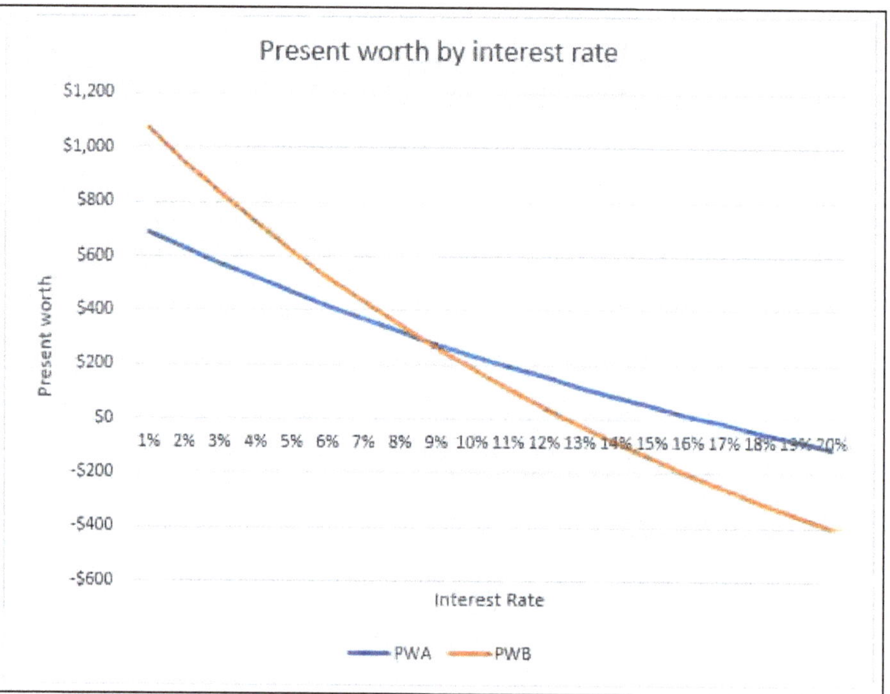

Fig. 4.9. PW value for option A and option B for interest rates 1% to 20% in Example 4.8.

Using MS Excel to Solve Chapter Examples

As in previous, chapters the following section presents the solutions of some of the previous problems using Excel.

Example 4.1: Excel solution

Michael just bought a car for $22,000. Michael estimates that the car operation and maintenance cost will be $250 per month. He believes that he will keep the car for six years and that after using the car he will be able to resell it for $4,000. Consider a 6% yearly interest rate. Draw the cash flow and determine the present worth for his car:

Solution: The solution formula is shown in the f_x rectangle, where the present worth is calculated by adding the initial investment amount and the present worth calculation using the present value formula PV and plugging in the values for monthly rate, the number of months, and the salvage value of $4,000 at the end of the 72 periods.

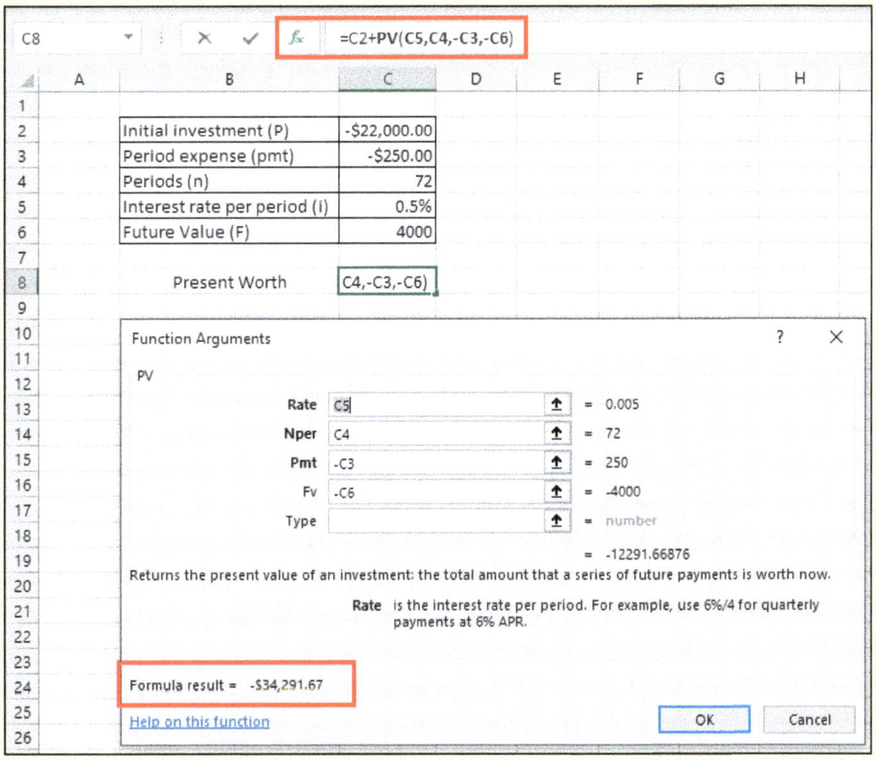

Fig. 4.10. Example 4.1 Excel solution.

Example 4.2: Excel Solution

Tom, Michael's brother, bought an e-car for $33,000. He received an instant rebate of $2,000 from New York state. Unfortunately, he spent $2,500 setting up a charging station at home, but car operation and maintenance costs are estimated to be only $40 per month. Tom is also keeping the car for six years, and he thinks that he will be able to resell it for $9,000. Consider a 6% yearly interest rate.

Solution: The solution formula is shown in the f_x rectangle; in Example 4.11 the present worth is calculated by adding the initial investment amount and the present worth calculation using the present value formula PV.

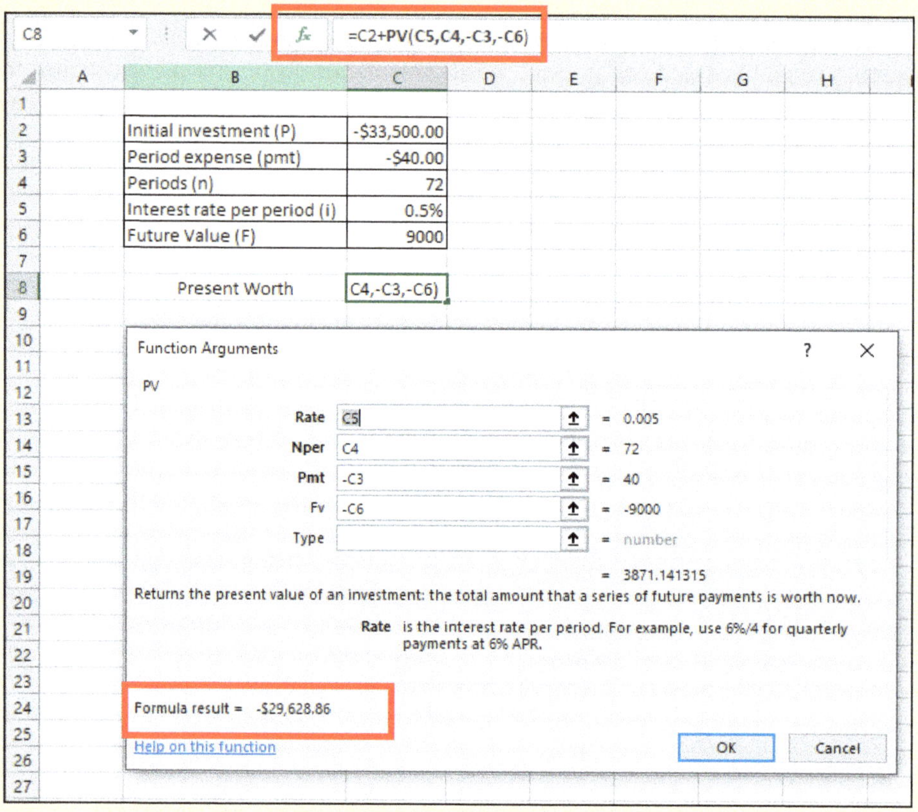

Fig. 4.11. Example 4.2 Excel solution.

Example 4.3: Excel Solution

Paula will be renting a nice two-bedroom apartment for a monthly payment of $1,500. Her lease stipulates that her rent will increase by 3% each year. Determine the PV for renting after keeping the apartment for 3 years, 10 years, and 20 years. Consider a 4% yearly interest rate.

C9 =NPV(4%,C5:C7)

For three years

Period	Amount per period
0	$0
1	-$18,000
2	-$18,540
3	-$19,096
PV=	($51,425.42)

For ten years

Period	Amount per period
0	$0
1	-$18,000
2	-$18,540
3	-$19,096
4	-$19,669
5	-$20,259
6	-$20,867
7	-$21,493
8	-$22,138
9	-$22,802
10	-$23,486
PV=	($165,776.85)

For twenty years

Period	Amount per period
0	$0
1	-$18,000
2	-$18,540
3	-$19,096
4	-$19,669
5	-$20,259
6	-$20,867
7	-$21,493
8	-$22,138
9	-$22,802
10	-$23,486
11	-$24,190
12	-$24,916
13	-$25,664
14	-$26,434
15	-$27,227
16	-$28,043
17	-$28,885
18	-$29,751
19	-$30,644
20	-$31,563
PV=	($316,285.94)

Fig. 4.12. Present worth Excel solution for Example 4.3.

The NPV formula considers the cash flow from periods 1 until n = 3, n = 10 and n = 20.

Consider that instead of renting Paula could buy a house with a $40,000 down payment and a $1,550 monthly payment, which includes property taxes, maintenance costs, and a mortgage payment.

Determine the PV after keeping the house for 3 years, 10 years, and 20 years. Assume that the net flow after reselling the house and paying what is left of the mortgage is $24,000 after 3 years, $114,000 after 10 years, and $339,000 after 20 years. Assume a 4% yearly interest rate.

Example 4.3: Excel Solution (Continued)

C9 =NPV(4%,C5:C7)+C4

For three years

Period	Amount per period
0	-$40,000
1	-$18,600
2	-$18,600
3	$5,400
PV=	($70,280.78)

For ten years

Period	Amount per period
0	-$40,000
1	-$18,600
2	-$18,600
3	-$18,600
4	-$18,600
5	-$18,600
6	-$18,600
7	-$18,600
8	-$18,600
9	-$18,600
10	$95,400
PV=	($113,848.35)

For twenty years

Period	Amount per period
0	-$40,000
1	-$18,600
2	-$18,600
3	-$18,600
4	-$18,600
5	-$18,600
6	-$18,600
7	-$18,600
8	-$18,600
9	-$18,600
10	-$18,600
11	-$18,600
12	-$18,600
13	-$18,600
14	-$18,600
15	-$18,600
16	-$18,600
17	-$18,600
18	-$18,600
19	-$18,600
20	$320,400
PV=	($138,064.90)

Fig. 4.13. Present worth Excel solution for Example 4.3.

Example 4.4: Excel Solution

Jim invested $5,000 in a small business and received $1,000 in years 1, 2, and 3. In year 4 there was a fire, so he did not receive money from his investment in years 4 and 5. The business reopened and in year 6 he received benefits for $2,000. In year 7 he received benefits for $1,500, in year 8 he received benefits for $1,000, in year 9 he received benefits for $500, and in year 10 he did not receive any benefits, so they decided to close the business. Determine the PV for his investment with a yearly interest rate of 6%.

Solution: The solution formula is shown in the f_x rectangle; in the previous examples the present worth was calculated by adding the initial investment amount and the present worth calculation using the net present value formula NPV for the cash flow.

Example 4.4: Excel Solution (Continued)

C14 =C3+NPV(6%,C4:C12)

Periods	Cash flow
0	-$5,000
1	$1,000
2	$1,000
3	$1,000
4	$0
5	$0
6	$2,000
7	$1,500
8	$1,000
9	$500
	$1,003.88

Fig. 4.14. Present worth MS Excel solution for Example 4.4.

Example 4.5: Excel Solution

Determine the present worth for the following two options, use a yearly interest rate of 6%.

Option	Cost	Service Life in Years	Savings Per Year	Salvage Value
A	$4,484	3	$1,800	0
B	$7,985	6	$2,700	$1,500

Solution: As in previous examples the present worth is calculated by adding the initial investment amount and the net present worth calculation using the net present value formula NPV. This same solution approach can be applied to Examples 4.6 and 4.8.

C11 =C3+NPV(6%,C4:C9)

Periods	Option A Cash flow	Option B Cash flow
0	-$4,484	-$7,985
1	$1,800	$2,700
2	$1,800	$2,700
3	-$2,684	$2,700
4	$1,800	$2,700
5	$1,800	$2,700
6	$1,800	$4,200
	$602.33	$6,349.22

Fig. 4.15. Solution for Example 4.5.

Example 4.6: Excel Solution

Determine the present worth for the following two options, use a yearly interest rate of 6%.

Option	Cost	Service Life in Years	Savings Per Year	Salvage Value After 3 Years
A	$4,484	3	$1,800	0
B	$7,985	6	$2,700	$2,000

Solve this problem considering a life of 3 years. The solution considers the PV function instead of the NPV function described in the previous example.

D9 =PV(D6,D5,-D4,-D7,)+D3

	A	B	C	D
1				
2			Option A	Option B
3		Initial investment (P)	-$4,484.00	-$7,985.00
4		Savings per year (pmt)	$1,800.00	$2,700.00
5		Periods (n)	3	3
6		Interest rate (i)	6%	6%
7		Salvage Value	$0.00	$2,000.00
8				
9		Present Worth	$327.42	$911.37

Fig. 4.16. Solution for Example 4.6.

What Did We Learn in This Chapter?

1. Present worth is calculated by bringing the different cash flows from the future to an initial time; this time is frequently period 0.
2. Present worth can be used for selecting between different alternatives by choosing the alternative with the highest worth.
3. When calculating alternatives with different lives it is required to find the lowest common multiple for both life durations and then perform the financial analysis using the common multiple for both alternatives.
4. When calculating alternatives with different lives, it is assumed that the replacement's cost is identical to the original cost and that the option will have the same duration as originally stated.
5. When calculating alternatives with different lives, we can also adjust the duration for the project to the lowest life or to a different life value.
6. When calculating infinite periods, use the formula $P = \frac{A}{i}$.
7. When selecting between different alternatives the interest rate may affect the selection of the alternative.
8. The net present value formula, NPV, from Excel, can calculate a given cash flow from the first period to the last cash flow period.

Review Questions

Directions: Refer to what you learned in this chapter in order to answer the following questions:

1. What is the practical use of calculating a present value?
2. How would you use PV to determine the best investment option?
3. What is the difference between a present value calculated from a determine number of annuities and from an infinite number of annuities?
4. How many different types of gradients are there? Are there any calculation differences?
5. What is the approach to calculate an increasing gradient? Is this approach different from a decreasing gradient?
6. If you have an apartment lease, would you rather have the rent increase using a geometric or an arithmetic gradient?

Chapter 4 Practice Problems

Use the following problems to practice your newly acquired knowledge. Answers for even problems are in the back of the book.

1. What is the present worth of buying a machine with an initial cost of $10,000 that may save $2,000 per year for 10 years? Determine the present value for this equipment using a 5% yearly interest rate.

2. Determine the PW of a streaming service that costs $12.99 a month, assume that you need to pay the first installment in time zero and at the beginning of each month for the following 11 months. Consider a 7% yearly interest rate.

3. A new piece of equipment has a life of 15 years and requires an initial investment of $200,000. It has operation and maintenance costs at $20,000 per year. The equipment also requires a yearly license with an expected cost of $500 for the first year, and it is expected that this license cost will increase by $500 per year. Considering a yearly interest rate of 4%, determine the present worth of the investment.

4. A mechanical engineer is acquiring new equipment that has an initial cost of $50,000, a service life of 15 years and requires a yearly license that has cost of $8,000 for the first year, and that is scheduled to increase by 10% per year. Assuming a yearly interest rate of 4%, determine the present worth for the investment.

5. After completing his studies Jonah accumulated student loans for $100,000. His loan charges 18% interest per year.

   a. If he agreed to pay $2,000 each month, determine the number of monthly payments he would have to pay to cover the loan.

   b. Determine the value of his monthly payment if he pays his loan in 10 years.

6. A construction business requires a water pump for their operations. According to their estimate the current water pump has a yearly maintenance cost of $35,000 and it could last for 7 more years. Consider this to be option A.

The company could also replace the old water pump by acquiring a new water pump for $125,000. The new equipment would have a yearly maintenance cost of $7,500 and would last for 7 years with a no salvage value. Consider this to be option B. Determine if they should keep or replace their current water pump. Use a 7% interest rate per year.

7. Jonah has two equipment options; use PW to determine which option should be selected. Assume a 5% interest rate per year.
   - Option 1 has an initial cost of $650, delivers yearly benefits of $250, and has a duration of three years.
   - Option 2 has an initial cost of $950, delivers yearly benefits of $400, and has a duration of four years.

8. Calculate the PW for a computer that will be used in a 9-year project. The computer needs to be replaced every 3 years. Consider that the computer has a cost of $4,500 and has a salvage value of $700. In addition, the computer software needs to be updated each year with a cost of $1000. Use a 7% interest rate per year.

9. Jimmy saved $350 each month for 35 years for his retirement. Determine the value of his savings at the end of the 35 years. Considering your future value calculations, determine the amount that he could withdraw monthly for the next 20 years. Also, determine how much he could withdraw each month indefinitely. Use a 9% interest rate per year.

10. An animation company knows that they need to renew their computer equipment every two years to avoid obsolescence. Therefore, every two years they invest $50,000 in new hardware, which they resell for at the end of its useful life for 25% of the purchasing cost. Determine how much they should invest today in mutual funds with monthly compounding in order to have enough funds to replace this hardware indefinitely. Consider a 10% interest rate per year.

11. Matt just graduated college with a computer science degree. According to his calculations he will earn $90,000 in his first year and believes that his salary will increase by 15% each year. If Matt invested $250,000 in obtaining his degree, determine the PW for the next 10 years considering a 5% interest rate.

12. Determine the PW for the following cash flow; use a 7% yearly interest rate:

Periods	Cash Flow per Year
0	-$15,000
1	$2,000
2	$3,000
3	$4,000
4	$5,000
5	$4,000
6	$3,000
7	$2,000
8	$3,000
9	$4,000
10	$5,000

13. Jimmy is considering buying a 3D printer, and his options are as follows. Using a 5% interest rate per year, determine the present worth and which option he should choose.

Periods	Option A Cash Flow	Option B Cash Flow
0	-$15,000	-$20,000
1	$2,000	$6,000
2	$2,000	$5,000
3	$2,000	$4,000
4	$2,000	$3,000
5	$2,000	$2,000
6	$2,250	$2,000
7	$2,500	$2,000
8	$2,750	$2,000
9	$3,000	$2,000
10	$3,250	$2,000

14. George invested in a small business; the following table describes the business cash flow, determine PW for his investment assuming a 15% yearly interest rate.

Periods	Cash Flow per Year
0	-$150,000
1	$20,000
2	$25,000
3	$30,000
4	$35,000
5	$0
6	$0
7	$50,000
8	$50,000
9	$40,000
10	$50,000

15. Using PW determine if you should pay your car insurance in a lump sum of $2,500 at time zero, $700 four times a year, or $265 per month. Use a 4% interest rate

16. Using PW, determine the value of P for the following cash flow; assume an interest rate per period of 6%:

Periods	Cash Flow per Year
0	-$150,000
1	
2	
3	P
4	P+$10,000
5	P+$20,000
6	P+$30,000

Chapter 5 Equivalent Uniform Annual Flows Method

Paula's Copier Replacement Story

IMG 5.1

Once settled in her new job location, Paula noticed that people in her new office had problems using the copier. The equipment was old, the quality of the copies was poor, and paper jams were frequent.

In one of her first meetings with her new boss, Paula mentioned what she had been seeing with the copy machine; her boss said that she knew about the problem but had been too busy to find an appropriate replacement.

Therefore, Paula now had a new task. Her boss asked her to find a replacement, so now she would have to research different options and determine which copier should be acquired. Although it seemed easy, the task was more difficult than she anticipated. First, she had to find different options and then narrow it down to the best two. In addition, her analysis not only needed to include equipment cost, but also operation cost, maintenance cost, equipment life, and salvage value.

After a couple of weeks, Paula determined the best two options were brands A and B. The information for these options is described in the following table:

Option	Cost	Service Life in Years	Monthly Operation Cost	Salvage Value
A	$4,484	3	$150	0
B	$7,985	6	$225	$500

As a last step in the decision-making process Paula's boss asked her to perform the financial analysis for both options to support her selection. She was told to consider a yearly interest rate of 3% and to determine the monthly cost, as this cost would have to be included in the monthly operation budget.

Reading Activity: Responding to Paula's Copier Story
1. What is the difference between a PW analysis and a monthly analysis?
2. Should the time horizon be 3 years or 6 years?
3. Would there be any advantages to determining the monthly cost instead of the present worth?
4. How should nonuniform costs be considered? For example, the copier may require an additional unscheduled service.

Learning Outcomes

After covering this chapter, the student should be able to do the following:

1. Learn how to determine equivalent uniform annual flows when considering cashflows that include benefits and costs for one or more mutually exclusive options.
2. Learn how to calculate equivalent uniform annual flows (EUAF) that only consider capital costs.
3. Learn how to determine equivalent uniform annual flows (EUAF) for infinite durations.
4. Learn how to reduce the number of loan payments by increasing the payment amount.
5. Learn the basics of personal finance, including what the FICO score is.

Introduction: Comparison Methods

In the previous chapter we presented a way to compare different cash flows by bringing all cash flows to a present value to select an option. In this chapter we consider the method of Equivalent uniform annual flows (EUAF), which transforms a cash flow into a series of uniform annual flows. Performing correct present worth or equivalent uniform worth calculations will determine the best option among different options.

The introductory reading describes situations where knowing which option has a lower cost using a present value may not provide the necessary information. Often, a person or company needs to estimate the uniform monthly benefits or costs of certain items. For example, in personal finance it is useful to verify if a person can afford a monthly mortgage based on their income. These uniform flows are defined as annuities, the word *annuity* may be a bit deceiving as these uniform flows may correspond to different time periods such as weeks, months, quarters, or semesters, and not necessarily years.

In addition, there are different equivalent uniform annual flows (EUAF), such as equivalent uniform annual worth (EUAW) in which projects include positive and negative flows and where the highest EUAW positive values should be selected when analyzing several projects. Similarly, the Equivalent Uniform Annual Benefits (EUAB) only considers benefits, in other words positive income; therefore, for this case the objective is the same as for the case of EUAW and the highest EUAB positive value should be selected. In contrast, for the case of equivalent uniform annual cost (EUAC) these cash flows only consider costs and the lowest EUAC positive values should be selected when there are multiple options. It is important to note EUAC is described using positive values, so negative EUACs represent negative cost or, in other words, an income.

Equivalent Uniform Annual Costs for Capital Costs and Annuities

The following examples show how to calculate equivalent uniform annual costs for the first two copier options described in the initial reading.

Example 5.1

From the information provided in the introduction, alternative A has an initial cost of \$4,484, a life of three years, a monthly operation cost of \$150, and a yearly interest rate of 3%. Determine the equivalent uniform annual cost (EUAC).

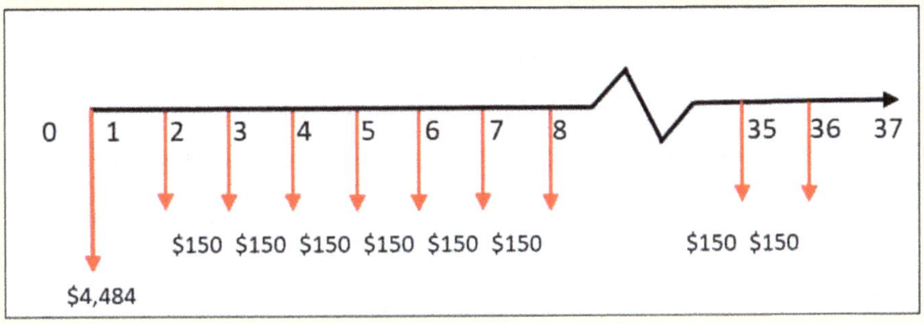

Fig. 5.1. Example 5.1 cash flow: option A.

First, let us calculate the interest rate per month. As before, this is done by dividing the yearly interest rate by the number of months in a year:

$$i = \frac{3\%}{12} = 0.25\%$$

The monthly cost of the device A is calculated by calculating the annuity from the initial cost. Using annuity from a present.

$$A = \$4,484(A/P, 0.25\%, 36) = \$4,484 * 0.0291 = \$130.48$$

Since the monthly operation cost of \$150 is already in months, no calculation is needed. Therefore, the monthly cost is obtained by adding the cost from acquiring the equipment and the operation cost:

$$\text{Total monthly cost} = \$150 + \$130.48 = \$280.48$$

Similarly, for option B, determine the (EUAW) assuming a life of six years (72 months), an initial cost of \$7,985, monthly operation costs of \$225, and a salvage value of \$500 after six years. Figure 5.2 shows the cash flow for this alternative.

Example 5.1 (Continued)

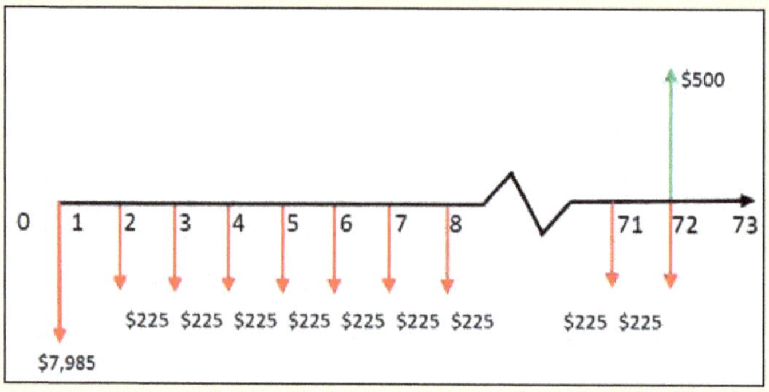

Fig. 5.2. Example 5.2 cash flow: option B.

Just like option A the interest rate per period is 0.25%. The monthly cost is now $225, and the monthly cost of device B is calculated as the annuity from a present value: $7,985 (A/P, 0.25%,72).

$$\$7{,}985 * 0.0152 = 121.37$$

Since the monthly operation cost of $225 is already in months, no calculation is needed. Therefore, the monthly cost is obtained by adding the costs from buying the copier and from the operation costs:

$$\text{Total monthly cost} = \$225 + \$121.37 = \$346.37$$

On the other hand, the salvage value is also brought to annuities: $500 (A/F, 0.25%,72)

$$A = \$500\ (A/F, 0.25\%, 72) = \$500 * 0.0127 = \$6.35 \text{ (Salvage value is a benefit)}$$

Therefore, the total monthly cost is the initial investment: the monthly costs minus the salvage value from the future.

$$A = \$346.37 - \$6.35 = \$340.02$$

Note that cost values are represented using positive numbers.

From Example 5.1 we learned the following:

1. (EUAW) considers positive flows (benefits / savings) and negative flows (costs) and benefits to determine annual worth.
2. Equivalent Uniform Annual Cost (EUAC) series only considers costs and are generally expressed using positive values.
3. For EUAC cost values are positive for a disbursement of money; negative values mean a positive inflow of money.
4. Salvage values occur at the end of the life of an item, and they are often positive. However, this is not always the case as the disposal of some products can generate additional expenses.

Exercise

Determine the (EUAC) from buying a new car that has an initial cost of $30,000 and a yearly operation cost of $1,500. Consider a yearly interest rate of 3% and a life of 15 years.

Answer: EUAC = $4,013.00

Different Lives Analysis

Chapter 4 describes situations where alternatives have different lives. In such cases it is required to determine the lowest common multiple for both life durations and then use the lowest common multiple to perform the financial analysis for both alternatives. For example, if one alternative has a life of three years and the other alternative has a life of two years, then both alternatives are analyzed for a period of six years.

On the other hand, for (EUAF), when the duration in one of the alternatives is the multiple of the other, we can conduct the analysis using the original life durations. To clarify this point let us perform an analysis for option A, considering six years instead of the original three years. This is done because the other option used for comparison, option B, has a life of six years and is a multiplier of three years, the original life duration for option A. This is described in Example 5.2.

Example 5.2

Determine the EUAC, using the information provided for option A and assuming a life of six years. Option A has an initial cost of $4,484, a monthly operation cost of $150, and a yearly interest rate of 3%.

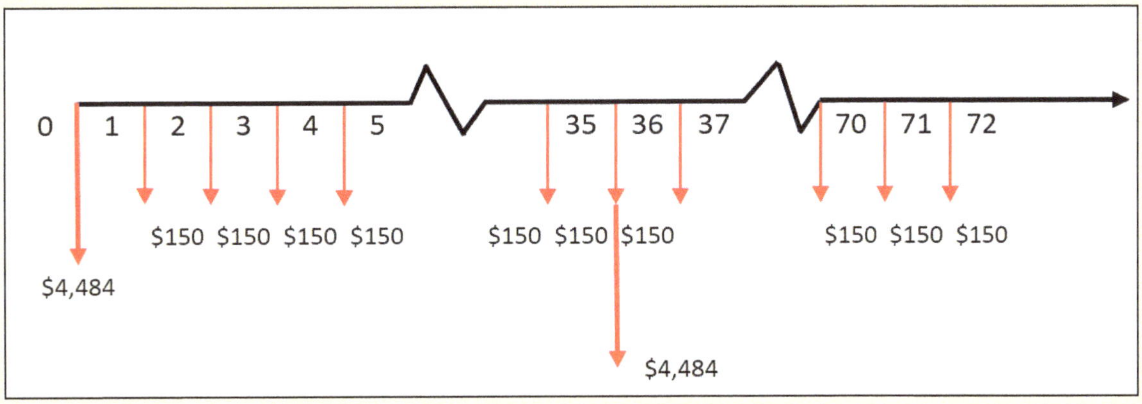

Fig. 5.3. Example 5.2 cash flow for option A for 72 months.

Assume an interest rate of 0.25% rate per month.

First, the initial purchase will be brought to a monthly cost using the following equation.

$$\$4{,}484(A/P, 0.25, 72) = \$4{,}484 * 0.0152 = \$68.15$$

Example 5.2 (Continued)

Second, the replacement purchase will be brought to a monthly cost by completing two steps:

1. Bringing the replacement cost to the present in time 0
2. Spreading this cost into the total duration of the flow for 72 months using the following equation:

A = $4,484 * (P/F, 0.25%, 36) * (A/P, 0.25%,72) = $4,484 * 0.9140 * 0.0152 = $62.29

Note that these two factors are multiplied.

As in previous examples we add the monthly cost $150 to determine the total cost per period. Since these periods are in months, they are the total monthly cost.

Total monthly cost = $68.15 + $62.29 + $150 = $280.44

Let us now compare the results obtained for option A from Example 5.1 and Example 5.2. Option A has a monthly cost of $280.48 for a life of three years and six years. Therefore, it is important to note that an option maintains its cost for life durations that are a multiple of the original duration. As observed in Examples 5.1 and 5.2, the cost of option A remains constant for a duration of three and six years.

This is because it is assumed that the replacement maintains its cost regardless of the time difference. However, this would not occur if the replacement had a different cost or if the value of the life of the item was modified from the original assumption, for example a life of five years instead of life of six years.

We can also observe that the yearly cost difference of $0.04 between both option A solutions is due to a rounding error.

From Example 5.2 we learned the following:

1. For EUAC with multiple options only one option can be selected; the best option is the one with the lowest positive cost.
2. With options with different lives and when one of them has a duration that is a multiple of the other, it is not required to calculate the EUAC for the duration of the other if and only if we have an identical replacement and equal duration for the replacement.

Exercise

Verify the yearly cost for option B for 12 and 18 years. Assume identical replacement.

Capital Cost Analysis

There are situations in which it is desired to calculate the equivalent uniform cost when only the initial cost and salvage value are considered. For example, consider buying a car where only the car cost and the resale values are known, the monthly operation and maintenance cost are unknown, or it is not required to consider these costs. Equation 5.1 describes how to determine the equivalent uniform cost using capital cost.

Equation 5.1 Equivalent uniform cost from capital cost

$$A = P(A/P, i, n) - S(A/F, i, n)$$

Where A is an annuity, P is the present value, and S is the salvage value. Note that since Equation 5.1 is used to calculate costs, if the result is positive, it indicates a positive cost, and if the result is negative, it indicates a benefit. This is shown in Example 5.3.

Example 5.3

Dan, a graphic designer, needs to renew his computer equipment to avoid obsolescence. Since Dan's computers are high-end, they are still attractive to other users. Assume that Dan invests $6,000 in new equipment every two years and later sells this equipment for $1,500. Determine the monthly cost from buying new equipment. Use an interest rate of 6% per year.

Present cost = $6,000 | Salvage value = $1,500

Interest rate $i = 6\%$ per year $i = 0.5\%$ per month | n = 24 months

A = monthly cost

$$A = P(A/P, i, n) - S(A/F, i, n)$$

$$A = \$6{,}000(A/P, 0.5\%, 24) - \$1{,}500(A/F, 0.5\%, 24)$$

$$A = (\$6{,}000 * 0.0443) - (\$1{,}500 * 0.0393)$$

$$A = (\$265.8) - (\$58.95) = \$206.85$$

Exercise

Determine the EUAC for a new machine that has an initial cost of $150,000 and a salvage value of $15,000 after 10 years of use. Consider a yearly interest rate of 7%.

Answer: EUAC = $20,270.96

Annuities and Gradients

Chapter 3 introduced the concept of annuities and how to calculate annuities from a gradient using Equation 3.18. Example 5.4 describes how to apply Equation 3.18 for the calculation of equivalent uniform flows from a gradient.

Equation 3.18: Annuity from a gradient

$$A = G\left[\frac{(1+i)^n - in - 1}{[i(1+i)^n - i]}\right]$$

Example 5.4

Mike is investing in a new project in which he plans to invest $1,000 next month and then increase his monthly investment by $1,000 for the next 72 months.

Therefore, his investment for month 1 is $1,000, $2,000 for month 2, $3,000 for month 3, and so on until month 72. Assume an interest rate of 12% per year. However, Mike wants to know how much he would have to invest monthly at a constant rate to reach the same amount of money if he were increasing his monthly amount as described. In other words, calculate the annuity for Mike's cash flow.

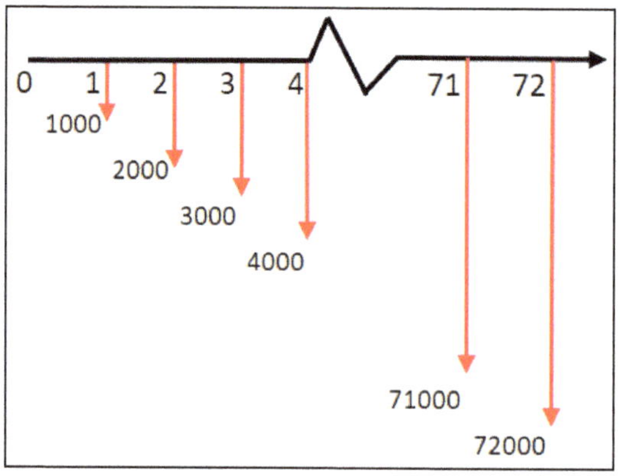

Fig. 5.4. Example 5.4 cash flow.

The solution considers two components as shown in Fig. 5.5. One is the constant amount from the annuities, and two is the constant amount from the gradient. Therefore, the constant amount from the annuity has a value of $1,000, and the constant amount from the gradient is calculated by applying Equation 3.18.

Example 5.4 (Continued)

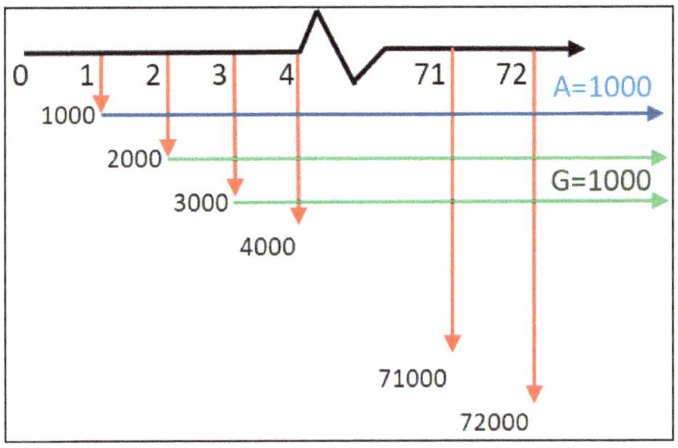

Fig. 5.5. Example 5.4 cash flow showing the constant annuity amount and the gradient amount.

Using Equation 3.18 we can determine the annuity value from a gradient:

$$A = G\left[\frac{(1+i)^n - in - 1}{[i(1+i)^n - i]}\right]$$

$$G = \$1000$$

$$A = \$1{,}000\left[\frac{(1+0.01)^{72} - (0.01 * 72) - 1}{[0.01(1+0.01)^{72} - 0.01]}\right] = \$1{,}000\left[\frac{0.3271}{0.0105}\right]$$

$$A = \$1{,}000 * 31.2386 = \$31{,}238.6$$

Therefore, the annuity value is calculated by adding the constant annuity $A = \$1{,}000$ and the annuity from the gradient $A = \$31{,}238.6$:

$$A = \$1{,}000 + \$31{,}238.6 = \$32{,}238.6$$

Exercise

Russell just bought a house for \$450,000 and is planning to live in it for 20 years, after which he expects to sell the house for \$850,000. To sell the house at a competitive price he needs to spend \$1,500 in maintenance after the first year, and he thinks that the maintenance cost will increase by \$150 each year, so in the second year it will be \$1,650. Determine the EUAC assuming a life for the house of 20 years and a yearly interest rate of 5%. Hint: Do not consider the salvage value.

Answer:

PV of Cost= \$483,466.58

EUAC = \$38,794.61

Annuities for Project Selection

The introductory reading presented the selection between two options as described in Example 5.1. Similarly, it is possible to use annuities to calculate benefits or to apply annuities for the selection of the best option among more than two options. Example 5.5 considers a benefit project selection problem for 4 options.

Example 5.5

Dan is a lucky lottery winner. He got an after-tax prize of $2,500,000. As soon as he received the money, he started receiving multiple phone calls from people offering investment opportunities. From the different offers he received he put the four options that sounded attractive in the following table.

Dan would like to invest in a project that will deliver the highest monthly benefit and is considering an interest rate of 9% per year for a life of 10 years for all options. Assume that the salvage value occurs at the last period.

Table 5.1: Data for Example 5.5

Option	A	B	C	D
Initial investment	$500,000	$750,000	$1,200,000	$350,000
Monthly operation cost	$25,000	$50,000	$80,000	$5,000
Monthly benefit	$40,000	$65,000	$110,000	$25,000
Salvage value	$600,000	$1,000,000	$1,000,000	$600,000

According to the data the project is 10 years long. Therefore $n = 10 * 12 = 120\ months$

And the monthly interest rate per year is $i = {}^{9\%}/_{12} = 0.75\%\ per\ month$

Considering the investment and the monthly costs as negative and the salvage value as positive, the equation for option A is as follows:

$$\text{EUAW}_A = -\$500{,}000\ ({}^{A}/_{P}, 0.75\%, 120) + \$40{,}000 - \$25{,}000 + \$600{,}000\ ({}^{A}/_{F}, 0.75\%, 120)$$

$$\text{EUAW}_A = (-\$500{,}000 * 0.0127) + \$40{,}000 - \$25{,}000 + (\$600{,}000 * 0.00517)$$

$$\text{EUAW}_A = (-\$6{,}350.00 + \$15{,}000 + \$3{,}102.00) = \$11{,}752.00$$

Option B is as follows:

$$\text{EUAW}_B = -\$750{,}000\ ({}^{A}/_{P}, 0.75\%, 120) + \$65{,}000 - \$50{,}000 + \$1{,}000{,}000\ ({}^{A}/_{F}, 0.75\%, 120)$$

$$\text{EUAW}_B = (-\$750{,}000 * 0.0127) + \$65{,}000 - \$50{,}000 + (\$1{,}000{,}000 * 0.00517)$$

$$\text{EUAW}_B = (-\$9{,}525.00 + \$15{,}000 + \$5{,}170.00) = \$10{,}645.00$$

Example 5.5 (Continued)

Option C is as follows:

$$EUAW_C = -\$1{,}200{,}000\left(A/_P, 0.75\%, 120\right) + \$110{,}000 - \$80{,}000 + \$1{,}000{,}000\left(A/_F, 0.75\%, 120\right)$$

$$EUAW_C = (-\$1{,}200{,}000 * 0.0127) + \$110{,}000 - \$80{,}000 + (\$1{,}000{,}000 * 0.00517)$$

$$EUAW_C = (-\$15{,}240.00 + \$30{,}000 + \$5{,}170.00) = \$19{,}930.00$$

Option D is as follows:

$$EUAW_D = -\$350{,}000\left(A/_P, 0.75\%, 120\right) + \$25{,}000 - \$5{,}000 + \$600{,}000\left(A/_F, 0.75\%, 120\right)$$

$$EUAW_D = (-\$350{,}000 * 0.0127) + \$25{,}000 - \$5{,}000 + (\$600{,}000 * 0.00517)$$

$$EUAW_D = (-\$4{,}445.00 + \$20{,}000 + \$3{,}102.00) = \$18{,}657.00$$

Based on these calculations the results are as follows:

$$EUAW_A = \$11{,}752.00$$

$$EUAW_B = \$10{,}645.00$$

$$EUAW_C = \$19{,}930.00$$

$$EUAW_D = \$18{,}657.00$$

Therefore, the final selection is alternative **C** because it provides the highest monthly benefit.

Exercise

Determine the option that will provide the highest yearly benefit for the options described in the following table. Consider a time horizon of seven years for all alternatives and a yearly interest rate of 10% per year.

Option	A	B	C	D
Initial investment	\$5,000	\$7,500	\$10,000	\$13,000
Monthly Benefit	\$400	\$650	\$800	\$1,250

Answer:

A = \$373 B = \$109 C = \$146 D = \$80

Annuities for Loan Analysis

As described in this and previous chapters, we can calculate an annuity from a present and a present from an annuity using Equations 3.8 and 3.9.

Equation 3.8: Annuity from a present value

$$A = \frac{P(1+i)^n * i}{[(1+i)^n - 1]}$$

Equation 3.9: Present value from an annuity

$$P = \frac{A * [(1+i)^n - 1]}{(1+i)^n * i}$$

These two equations are frequently used to determine a loan payment amount. For instance, when buying a $25,000 car we may determine the monthly payment given a certain number of payments and an interest rate. However, we can also use these formulas to determine a balance on a loan after a certain number of payments. This is described in Example 5.6.

Example 5.6

Dan, the lucky lottery winner from Example 5.5, bought a house three years prior to winning his lottery prize. The house he bought had an original price of $550,000. He obtained a 30-year mortgage loan to acquire this house and agreed to make monthly payments at an interest rate of 6% per year.

Assume that at the time of winning the lottery Dan made a total of 38 monthly payments. Determine the following:

1. His monthly mortgage payment?
2. Assuming that Dan wants to make a full payment of the house, how much does he owe after 38 payments?

The data and solution for the first question is:

$P = \$550{,}000$

$i = {}^{6\%}/_{12} = 0.5\%\ per\ month$

$n = 30 * 12 = 360\ months$

Equation 3.8,

$$A = \frac{P(1+i)^n * i}{[(1+i)^n - 1]} = \frac{\$550{,}000(1+0.005)^{360} * 0.5}{[(1+0.005)^{360} - 1]} = \$550{,}000 * 0.00600 = \$3{,}300$$

The data and solution for the second question is as follows:

Since Dan made 38 payments, his remaining payments are 360 − 38 = 322 payments left to fully pay off his mortgage loan. Therefore, we use Equation 3.9 to solve the amount of money he owes:

$$P = \frac{A * [(1+i)^n - 1]}{(1+i)^n * i} = \frac{\$3{,}300 * [(1+0.005)^{322} - 1]}{(1+0.005)^{322} * 0.005} = \$3{,}300 * 159.8618 = \$527{,}543.87$$

Example 5.6 is a good example of the effect of compound interest and how this affects our daily lives. First let us consider the following. If we multiply the value of the monthly annuity, \$3,300, by the number of remaining payments, 322, we get \$1,062,600. This amount is quite different from the present value amount obtained in Example 5.6. This is a clear indication of the effect of monthly compounding.

Exercise

Jonah bought a car for \$10,000. Determine the monthly payment if his loan is for 60 months at 3% per year. Determine his debt after making 12 payments.

Answer: Annuity \$179.69

Balance after for after 12 payments (n=48) \$8,118.02

Based on the results in Example 5.6 and the previous exercise, it is clear that having to pay interest results in a higher end value as opposed to not having to pay interest at all. It is also obvious that the higher the interest rate the more money is spent paying off the interest amount.

However, it is more common that someone who is in debt will not have the same experience as described in Example 5.6; not everyone will receive an influx of money such that they can pay off their debt in full. A more realistic expectation is that they may be able to make a small extra payment every now and then, which could be very beneficial to their finances. One way to observe this is by considering the number of initial payments that would be made and then how this number is affected by making an extra payment.

To determine the number of periods, consider Equations 3.8 and 3.9 and consider that the present value, the annuity, and interest rate is known. Therefore, equating Equations 3.8 and 3.9 we obtain the following:

Equation 5.2 Number of periods n from an annuity and a present value

$$n = \frac{ln\left[(1 - \frac{P * i}{A})^{-1}\right]}{\ln(1 + i)}$$

Example 5.7 presents the application of Equation 5.2 using the values from Example 5.6.

Example 5.7

Dan bought a house for \$550,000. His mortgage charges a yearly interest rate of 6%, and he makes payments of \$3,300. Determine how many payments he has to pay.

$P = \$550{,}000$

$i = {}^{6\%}/_{12} = 0.5\%\ per\ month$

$A = \$3{,}300\ per\ month$

$$n = \frac{ln\left[(1 - \frac{P * i}{A})^{-1}\right]}{\ln(1 + i)}$$

Example 5.7 (Continued)

$$n = \frac{ln\left[(1 - \frac{\$550{,}000 * 0.005}{\$3{,}300})^{-1}\right]}{\ln(1 + 0.005)} = \frac{ln\left[(1 - \frac{\$550{,}000 * 0.005}{\$3{,}300})^{-1}\right]}{\ln(1 + 0.005)}$$

$$\frac{ln[6]}{\ln(1.005)}$$

$$n = \frac{1.791759}{0.004987} = 359.2470 \approx 360\ monthly\ periods$$

Example 5.8 presents a different application for this formula.

Example 5.8

Mary, Dan's neighbor, also bought a house with the same payment information as Dan's house.

$P = \$550{,}000$
$A = \$3{,}300\ per\ months$
$i = 0.5\%\ per\ month$

However, after paying her mortgage for 24 months according to the loan agreement Mary switched jobs and, as a result, received a signing bonus of $7,500 with an increase in income. Therefore, she will be able to increase her mortgage payment up to 20%.

Determine the following:

1. What is her balance?

Since Mary made 24 payments, she will have 360-24 = 336 payments left to fully pay off her mortgage loan. Therefore, we can use Equation 3.9 to solve for the amount of money she owes.

$$P = \frac{A * [(1 + i)^n - 1]}{(1 + i)^n * i}$$

$$P = \frac{A * [(1 + i)^n - 1]}{(1 + i)^n * i} = \frac{\$3{,}300 * [(1 + 0.005)^{336} - 1]}{(1 + 0.005)^{336} * 0.005}$$

$$P = \$3{,}300 * 162.5688 = \$536{,}477.1836$$

Example 5.8 (Continued)

2. How many more months she has left on her loan if she applies her signing bonus to her debt?

Consider a deduction of $7,500 from the current balance described in part 1; this becomes the new present value. Then, solve for the number of periods using Equation 5.2:

$$P = \$536{,}477.1836 - \$7{,}500 = \$528{,}977.1836$$

$$n = \frac{ln\left[(1 - \frac{P * i}{A})^{-1}\right]}{\ln(1+i)}$$

$$n = \frac{ln\left[(1 - \frac{\$528{,}977.1836 * 0.005}{\$3{,}300})^{-1}\right]}{\ln(1+0.005)} = \frac{ln\left[(1 - \frac{\$528977.1836 * 0.005}{\$3{,}300})^{-1}\right]}{\ln(1+0.005)}$$

$$n = \frac{ln[5.0372]}{\ln(1.005)}$$

$$n = \frac{1.616868}{0.004987} = 324.1814 \approx 324\ monthly\ periods$$

3. How many more months she has left on her loan if she increases her mortgage payment by 20%?

$$A = 3{,}300 * 1.2 = \$3{,}960$$

$$n = \frac{ln\left[(1 - \frac{P * i}{A})^{-1}\right]}{\ln(1+i)}$$

$$n = \frac{ln\left[\left(1 - \frac{\$536{,}477.1836 * 0.005}{\$3{,}960}\right)^{-1}\right]}{\ln(1+0.005)}$$

$$\frac{ln[3.09953]}{\ln(1.005)}$$

$$n = \frac{1.131249}{0.004987} = 226.8150 \approx 227\ monthly\ periods$$

Example 5.8 (Continued)

4. How many more months she has left on her loan if she applies her signing bonus and increases her mortgage payment by 20%?

$$P = \$528{,}977.1836$$

$$A = \$3{,}960$$

$$n = \frac{ln\left[(1 - \frac{P * i}{A})^{-1}\right]}{\ln(1 + i)}$$

$$n = \frac{ln\left[(1 - \frac{\$528{,}977.1836 * 0.005}{\$3{,}960})^{-1}\right]}{\ln(1 + 0.005)} = \frac{ln\left[(1 - \frac{\$528{,}977.1836 * 0.005}{\$3{,}960})^{-1}\right]}{\ln(1 + 0.005)}$$

$$\frac{ln[3.01115]}{\ln(1.005)}$$

$$n = \frac{1.102320}{0.004987} = 221.0148 \approx 221\ monthly\ periods$$

From Example 5.8 we learned the following:

1. Making a payment that goes towards paying the principal greatly reduces the number of payments.
2. The results show that the number of periods is reduced from 336 to 227 by increasing the annuity payments by 20%. This is due to the compounding effect of the annuities. Therefore, if we are interested either in generating wealth or reducing a debt balance, it is a good idea to increase payments in a constant manner.

Personal Finance and the FICO Score

It could be inferred from Examples 5.7 and 5.8 that borrowing money is negative and has negative consequences. In fact, that is not correct. Acquiring debt is one of the tools that people may use to create wealth and to improve their quality of life.

For example, through credit a person can buy a car for their daily commute to work or may be able to get student loans and in turn have access to a job. Similarly, credit allows us to afford supplies and equipment to create a business, so we may aspire to a better standard of living. Likewise, governments and companies benefit from low interest rates to build infrastructure or to expand their operations. Therefore, it should be considered as licit and fair to pay someone else a financial benefit in return for using their money for our benefit. On the other hand, as described in previous chapters, this must be done in an ethical way and without abusing others due to their lack of financial literacy.

FICO Score

The most used tool to determine credit worthiness for a person is their FICO credit score. This score was developed in 1956 by Bill Fair and Earl Isaac. The score uses historical credit reports to assign a measurement of credit worthiness and assigns score values that range from 300 to 850. Having a low score means that a person has an elevated risk level, and therefore, the interest rate that a person receives is likely substantially higher than the one offered to a person with a high FICO score.

FICO scores of less than 580 are considered poor (well below average), scores 581 to 669 are considered fair (lower than average), scores of 670 to 739 are considered good (average or slightly better than average), scores of 740 to 799 are considered very good (above average), and scores of 800 and above are considered exceptional (well above average).

Three credit organizations, Experian, Equifax, and TransUnion, gather *consumers'* credit use information and develop credit reports for lenders. When borrowers apply for a loan, loan lenders use the applicants' FICO score to determine their credit worthiness. Credit worthiness and FICO scores are then used to set the interest rate and the amount of the loan that could be awarded to the borrowers' application. According to FICO, 90% of related credit decisions made from financial institutions and businesses related to personal finance in the United States consider the use of FICO scores. Having poor or fair FICO scores could deny a person's access to credit or may substantially increase their interest rate when getting a loan.

FICO scores consider the following metrics:

1. Payment history, in other words if a person pays on time. This metric weighs 35% of the *borrower's* score.
2. Amounts owed, which relates to the ratio of current debt balance to available credit. This second metric weighs 30%.
3. Credit history, how long since a person opened an account. This metric represents 15%.
4. Credit mix, which relates to having different types of credit, such as personal or credit card loans, which are different. This metric weighs 10%.
5. New credit, which represents any new lines of credit or applying for a new credit. This metric also represents 10%.

Surprisingly, FICO scores are not affected by personal factors such as a person's age, income, or length of employment. Therefore, FICO scores are skewed towards older people and have a negative impact on the younger population as their length of credit record may be low. Moreover, since the FICO score is a metric that is based on the person's credit record, it dismisses valuable information that could positively benefit a person's credit worthiness, such as personal wealth. For example, a young person who just inherited $2,000,000 and has this money in a savings account, or who just inherited a house or other valuable asset, will still have a low FICO score because their credit record is most likely quite short. Similarly, when a person pays off a credit, such as finishing a mortgage, their FICO score decreases. The FICO score is also negatively affected when a hard credit inquiry is performed, such as when a person applies for a new credit or a new loan. In addition, some hard inquiries may be performed by other entities, such as a potential property owner. Soft credit inquiries could be performed by other noncredit entities such as a potential employer and do not negatively affect the FICO score.

How to Improve the FICO Score

Based on the FICO metrics a young person interested in generating a good FICO score should do the following:

1. Start early and open a small credit account; it could be from a bank, a credit card company, or a commercial credit entity such as a department store.
2. Pay the balance, if possible, at the end of each month.
3. Maintain the credit balance lower than 30% of the available credit.
4. Never miss a payment. If they are required to pay child support, make sure to never miss a payment.
5. Do not open too many credit accounts and do not frequently apply for credit.
6. When possible and if convenient, mix the types of credit. For instance, have a personal loan, or a car loan and a credit card loan.

A Word on Financial Advice.

Having a good FICO score is like having a healthy weight, it is a consequence of making correct decisions. There are many books about financial advice and the objective of this section is not to provide a compilation of the different books but to provide a few basic points that may help you develop a financial balance.

1. Develop a simple budget that may allow you to balance your income and your expenses. Although it is known that having a budget and keeping it is difficult, it is necessary to prepare a list of your fixed and recurring costs, such as rent, utilities, transportation, and allow for miscellaneous and unexpected expenses. Once the expense amount is calculated it should be contrasted with your income to determine your final balance at the end of each pay period. If your balance is negative or is just barely positive you will be in financial trouble regardless of how much you earn.
2. In connection with the previous point, you should aim to save between 10–20% of your income. One approach to achieve this is to reassess your current expenses and spend on what is needed instead of what is desired. For instance, having car insurance is needed, as it protects your finances. On the other hand, buying coffee from a coffee parlor may be desired but not be needed. One way to achieve this may be by finding ways to reduce the cost of desired items, for instance preparing coffee at home and using a thermos would provide the satisfaction of drinking coffee without generating the expense.
3. Have an expense plan. Use the information that you have from the future to your advantage. For example, planning your vacation ahead of time will help you find better travel deals.
4. Protect your wealth. Having insurance is not a way to generate wealth but to reduce your losses in case of misfortune.
5. Avoid paying interest when possible. This can be done in a variety of ways:
    a. First, as previously described, having a good FICO score will reduce the amount of money that can go to interest payments. So, try to have a FICO score above 740.

   b. Second, if you use your credit card to pay try to pay the full balance at the end of the month. If this is not possible, verify if there is a payment plan. If not, try to pay as much as possible to reduce the amount of interest payments. Also, give payment priority to credits with the highest interest rate.

   c. Third, do not fall for the balance transfer trap. Often people with a high credit card balance transfer their balance to a new credit card that offers a promotional 0% interest for a period of time. Even when this is a true offer, it frequently occurs that people transfer an amount that they cannot pay within the promotional period and since the new credit card charges high interest rates they end up paying a higher interest amount.

   d. Avoid borrowing from payday loans as they often charge very high interest rates.

2. In agreement with point 5, do not pay taxes if you do not have to. This point does not mean that you should break the tax law; it is suggesting you use an IRA or 401(k) to invest your savings. This way you will lower your taxable income and pay less taxes. On the other hand, you will pay taxes when you withdraw money from your savings.

3. Seek professional assistance for taxes and investment. Unless you are a certified accountant, you may not be aware of new tax regulations that may help you get a higher tax refund. Similarly, a fiduciary financial advisor puts your best interest above their own and has the ethical obligation to recommend the best investments for you. Moreover, when selecting investment plans look for diversified index funds and exchange-traded funds that charge lower fees.

4. If you are young, open a new credit line and use it wisely. Do not overspend.

5. Do not fall for the money-back trap. Apply for credit when needed, not when it is offered. Sometimes credit card companies will give you some money back if you spend a certain amount of money within a certain period. For instance, $150 dollars if you spend $500 in the next three months. The offer is correct, but it might create the habit of spending beyond your means, resulting in paying interest.

Using MS Excel to Solve Examples

Example 5.1: Excel Solution

From the information provided in the introduction, alternative A has an initial cost of $4,484, a life of three years, a monthly operation cost of $150, and a yearly interest rate of 3%. Determine the equivalent uniform annual cost (EUAC).

Similarly, for option B, determine the (EUAW) assuming a life of six years (72 months), an initial cost of $7,985, monthly operation costs of $225, and a salvage value of $500 after six years. Use a yearly interest rate of 3% per year. The cash flow for option B is shown below.

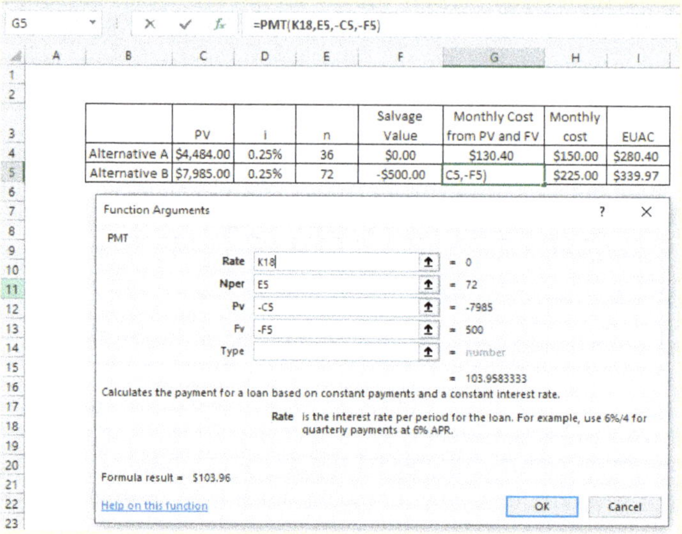

Fig. 5.6. Excel solution for problem 5.1

To find the monthly annuity value from the PV and the FV we use the excel function PMT. As shown in Fig. 5.6 the PMT function window requires the interest rate, the number of periods, the present value PV, and the future value FV. The total monthly cost is calculated by adding the calculated annuity from the PV and the FV with the monthly cost. Note that the number of periods for each alternative is different. Also observe that because Excel changes the sign for the PV and the FV, these values must be entered with an opposite sign in order to maintain their sign.

Example 5.2: Excel Solution

Determine the EUAC, using the information provided for option A and assuming a life of six years. Option A has an initial cost of $4,484, a monthly operation cost of $150, and a yearly interest rate of 3%.

	PV in year 0	PV for replacement month 36	i	n	Salvage Value	Monthly Cost from PV year 0	PV Cost from replacement year 0	Monthly Cost from PV year 0	Monthly cost	EUAC
Option A	$4,484.00	$4,484.00	0.25%	72	$0.00	$68.13	$4,098.53	$62.27	$150.00	$280.40
						=PMT(E4,F4,-C4,G4)	=PV(E4,36,,-D4)	=PMT(E4,F4,-I4)		=H4+J4+K4

Fig. 5.7. Excel solution for Example 5.2.

Example 5.2: Excel Solution (Continued)

In this solution the PV value from year 0 is brought to a monthly cost by using the function PMT for 72 months.

The replacement cost from month 36 is brought from the future to PV; this function is shown in cell I5. This value is then brought to a monthly cost using the function PMT for 72 months.

Finally, these two values are added to the monthly cost of $150, and the EUAC cost is calculated.

Example 5.3: Excel Solution

Dan, a graphic designer, needs to renew his computer equipment to avoid obsolescence. Since Dan's computers are high end, they are still attractive to other users. Assume that Dan invests $6,000 in new equipment every two years and later sells this equipment for $1,500. Determine the monthly cost from buying new equipment. Use an interest rate of 6% per year.

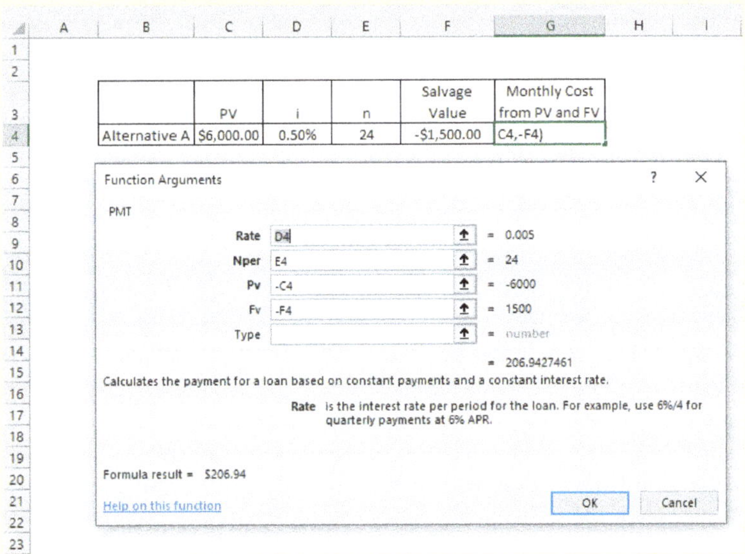

Fig. 5.8. Excel solution for Example 5.3.

As before we use the function PMT with opposite signs for PV and FV.

Example 5.4: Excel Solution

Mike is investing in a new project in which he plans to invest $1,000 next month and then increase his monthly investment by $1,000 for the next 72 months.

Therefore, his investment for month 1 is $1,000, $2,000 for month 2, $3,000 for month 3, and so on until month 72. Assume an interest rate of 12% per year. However, Mike wants to know how much he would have to invest monthly at a constant rate to reach the same amount of money if he were increasing his monthly amount as described. In other words, calculate the annuity for Mike's cash flow.

Example 5.4: Excel Solution (Continued)

	A	B	C	D
1				
2		Period		
3		0	0	
4		1	$1,000	
5		2	$2,000	
6		3	$3,000	
7		4	$4,000	
8		5	$5,000	
72		69	$69,000	
73		70	$70,000	
74		71	$71,000	
75		72	$72,000	
76			$1,649,017.73	=NPV(1%,C8:C79)
77			$32,238.61	=PMT(1%,B79,-C80)
78				

Fig. 5.9. Excel solution for Example 5.4.

Figure 5.9 shows a sequence starting from period 0 to period 72. Rows 6 to 68 are hidden but follow the described sequence. The formula in cell C76 is written in D76 and uses the NPV function to determine the PV for the 72-month flow for period 0. Once the PV for period 0 is calculated, the annuity is calculated using the PMT function that considers 72 periods and the NPV value previously calculated.

Example 5.5: Excel Solution

Dan is a lucky lottery winner. He got an after-tax prize of $2,500,000. As soon as he received the money, he started receiving numerous phone calls from people offering investment opportunities. From the different offers he received he put the four options that sounded attractive in the following table.

Dan would like to invest in a project that will deliver the highest monthly benefit and is considering an interest rate of 9% per year for a life of 10 years for all options. Assume that the salvage value occurs at the last period.

Table 5.1: Data for Example 5.5

Option	A	B	C	D
Initial investment	$500,000	$750,000	$1,200,000	$350,000
Monthly Operation Cost	$25,000	$50,000	$80,000	$5,000
Monthly Benefit	$40,000	$65,000	$110,000	$25,000
Salvage Value	$600,000	$1,000,000	$1,000,000	$600,000

Example 5.5: Excel Solution (Continued)

F7 =PMT(F6,F5,-F3,-F4)

	Alternative A	Alternative B	Alternative C	Alternative D	
Initial Investment	-$500,000.00	-$750,000.00	-$1,200,000.00	-$350,000.00	
Salvage Value	$600,000.00	$1,000,000.00	$1,000,000.00	$600,000.00	
Periods	120	120	120	120	
interest rate	0.0075	0.0075	0.0075	0.0075	
Annuity from investment and cost	-$3,233.24	-$4,333.11	-$10,033.52	-$1,333.11	=PMT(H68,H67,-H65,-H66)
Monthly cost	-$25,000.00	-$50,000.00	-$80,000.00	-$5,000.00	
Monthly Benefit	$40,000.00	$65,000.00	$110,000.00	$25,000.00	
EUAW	$11,766.76	$10,666.89	$19,966.48	$18,666.89	=H69+H70+H71

Fig. 10. Excel solution for Example 5.5.

This solution considers the annuity calculation using the values from the initial investment row 3 and the salvage value row 4 using the PMT function described in row 7. The EUAW is calculated from the sum of the annuity from the capital cost values row 7 and the monthly cost row 8 and monthly benefits row 9.

Example 5.6: Excel Solution

Dan, the lucky lottery winner from Example 5.5, bought a house three years prior to winning his lottery prize. The house he bought had an original price of $550,000. He obtained a 30-year mortgage loan to acquire this house and agreed to make monthly payments at an interest rate of 6% per year.

Assume that at the time of winning the lottery Dan made a total of 38 monthly payments. Determine the following:

1. His monthly mortgage payment?
2. Assuming that Dan wants to make a full payment of the house, how much he owes after 38 payments.

	PV	i	n	Monthly Payment	Formula
Question 1	$550,000.00	0.50%	360	$3,297.53	=PMT(D4,E4,-C4)
Question 2	$527,148.68	0.50%	322	$3,297.53	=PV(D5,E5,-F5)

Fig. 5.11. Excel solution for Example 5.6.

The solution for question 1 uses function PMT to find the annuity and the solution for question 2 uses the function PV to find the PV for the 322 remaining payments using the annuity value obtained in question 1.

Example 5.7: Excel Solution

Dan bought a house for $550,000. His mortgage charges a yearly interest rate of 6%, and he makes payments of $3,300. Determine how many payments he has to pay.

	PV	i	n	Monthly Payment	Formula
Question 1	$550,000.00	0.50%	359.247	$3,300.00	NPER(D3,-F3,C3)

Fig. 5.12. Excel solution for Example 5.7.

To find the number of periods we use the function NPER.

Example 5.8: Excel Solution

Mary, Dan's neighbor, also bought a house with the same payment information as Dan's house.

$P = \$550{,}000$
$A = \$3{,}300\ per\ months$
$i = 0.5\%\ per\ month$

However, after paying her mortgage for 24 months according to the loan agreement Mary switched jobs and, as a result, received a signing bonus of $7,500 with an increase in income. Therefore, she will be able to increase her mortgage payment up to 20%.

Determine the following:

1. What is her balance?
2. How many more months she has left on her loan if she applies her signing bonus to her debt?
3. How many more months she has left on her loan if she increases her mortgage payment by 20%?
4. How many more months she has left on her loan if she applies her signing bonus and increases her mortgage payment by 20%?

	PV	i	n	Monthly Payment	Formula
Question 1	-$536,477.18	0.50%	336	$3,300.00	=PV(D3,E3,F3)
Question 2	-$528,977.18	0.50%	324	$3,300.00	=NPER(D4,F4,C4)
Question 3	-$536,477.18	0.50%	227	$3,960.00	=NPER(D5,F5,C5)
Question 4	-$528,977.18	0.50%	221	$3,960.00	=NPER(D6,F6,C6)

Fig. 5.13. Excel solution for Example 5.8.

Figure 5.13 shows the functions used to solve this example using the NPER function.

What Did We Learn in This Chapter?

- Calculating equivalent uniform annual flows allows us to understand the cost per flow instead of the present value of cost.
- When considering flows with different lives, it is not necessary to use the common multiple if one of the options has a duration that is a multiple of the other if and only if there is an identical replacement and equal duration for the replacement.
- It is possible to learn EUAC by considering only capital costs.
- When selecting between different mutually exclusive options consider the option with the lowest EUAC.
- Annuities can be calculated from gradients using the factor A/G.
- For loan analysis, increasing the annuity amount modifies the number of payment periods.

Review Questions

1. Explain the difference between EUAB and EUAC?
2. What is a positive cost, and what is a negative cost?
3. How can you calculate EUAW from EUAB and EUAC?
4. How does the value of the annuity affect the number of periods?
5. Acquiring debt is always negative? Mention two instances when acquiring debt is a viable solution.
6. Is it a good business to pay a debt sooner if the debt has an interest rate that is below the inflation rate?
7. What is the FICO score, and how can you increase it?
8. What is the effect of having a good FICO score when getting a mortgage loan?

Chapter Formulas

Equation 3.8: Annuity from a present value

$$A = \frac{P(1+i)^n * i}{[(1+i)^n - 1]}$$

Equation 3.9: Present value from an annuity

$$P = \frac{A * [(1+i)^n - 1]}{(1+i)^n * i}$$

Equation 3.18: Annuity from a gradient

$$A = G\left[\frac{(1+i)^n - in - 1}{[i(1+i)^n - i]}\right]$$

Equation 5.2 Equivalent uniform cost from capital cost

$$A = P(A/P, i, n) - S(A/F, i, n)$$

Equation 5.2 Number of periods n from an annuity and a present value

$$n = \frac{ln\left[(1 - \frac{P * i}{A})^{-1}\right]}{\ln(1+i)}$$

Chapter 5 Practice Problems

Use the following problems to practice your newly acquired knowledge. Answers to even problems are in the back of the book.

Annuities

1. Nichole bought a $15,000 car, and after using it for five years she sold it for $2,500. Determine the EUAC assuming an interest rate of 7% per year. Hint: Only consider capital costs.

2. Sam is thinking about buying a new 3D printer. The printer costs $13,000 and requires a yearly maintenance of $2,000. Sam thinks that he may be able to earn $7,000 per year by providing 3D printing services to the local community. Consider that the printer may work for 10 years. Use a 10% yearly interest rate determine the following:

    a. EUAW with no salvage value

    b. EUAW with a salvage value of $3,500

3. Determine the EUAC for Sam's 3D printer in problem 2 considering only the capital costs.

4. Victoria wants to determine the EUAC for a 10 year period for a machine that had a cost of $25,000 and has a monthly cost of $250 for operation and maintenance. Furthermore, the machine requires a yearly service every 12 months with a cost of $1500. Assume a yearly interest rate of 6%.

5. Two machines are evaluated for a new project; consider a yearly interest rate of 9% and the information described in the following table. Determine EUAC for both options and determine the best option.

	Machine A	Machine B
Purchase cost	$10,000	$15,000
Salvage value	$1,250	$1,950
Duration	5 years	10 years

6. The local authorities are considering building a new bridge between two different zones of the city. An engineering firm is projecting the following bridge cost. Determine the EUAC for both options. Assume a yearly interest rate of 8%.

	Bridge A	Bridge B
Purchase cost	$100,000,000	$125,000,000
Duration	100 years	Infinite

7. Reevaluate Problem 5 with a duration of four years for machine A and a duration of twelve years for machine B.

8. Martin has the following options for acquiring a car for his ride-sharing services. Use EUAW with a yearly interest rate of 6%. Consider a time horizon of 4 years.

	Car A	Car B
Purchase cost	$50,000	$65,000
Maintenance cost per month	$2,500	$2,850
Benefit per month	$25,000	$32,500
Salvage value after 4 years	25 % from the purchase cost	35% from the purchase cost

Gradients

9. Nick invested $1,000 in his bank account today (year 0). He is planning on investing an amount (t) in year 1, and to continue with his investments in the following manner: invest (t + 150) in year 2, (t + 300) in year 3, (t + 450) in year 4 and to continue with this investment pattern for the following 25 years. Determine the equivalent uniform value for Nick's investment. Consider a yearly interest rate of 5%.

10. Josh bought an old car for $5,000, and unfortunately, he had to repair his car every 6 months. The first repair cost $250, the second repair cost $500, the third cost $750, and this pattern continued for the following repairs, until he sold the car after 10 repairs. Determine the PV for these repairs, assuming a yearly interest rate of 8%.

11. Solve problem 8 assuming that the monthly maintenance cost for car A increases by $20 per month while the monthly maintenance cost for car B decreases by $20 starting in month 1.

12. Determine the PV and the EUAC per year for Josh's car repairs from problem 10. Consider that in period 5, Josh had an extra expense of $4,000. Use a yearly interest rate of 8%.

13. Solve problem 8 assuming that the monthly maintenance cost for car A increases by $30 per month while the monthly maintenance cost for car B decreases by $10 starting in month 12.

Loan Analysis

14. Upon his graduation Albert accumulated a debt of 50,000 to obtain his degree. He is supposed to pay his loan in 10 years, making four payments per year at an interest rate of 8% per year. He will make his first payment one period after graduation. Determine the following:

    a. His loan payment per period
    b. His loan balance after 24 payments.
    c. The number of payments left after 24 payments if he increases his loan payment by 20%.

15. A company is considering acquiring office space; the space costs $4,000,000 the seller is willing to allow the first payment of $500,000 in six months after the time of purchase. The agreement also describes that the interest rate is 6% per year and that the company will make 36 monthly payments starting one month after the first payment.

    a. Determine the monthly payment amount.

    b. Determine the number of payments if the company doubles the payment amount after the 12 payment.

Chapter 6 Rate of Return and Incremental Analysis

Paula's Investment Story

IMG 6.1

IMG 6.2

After four years in her job, Paula's future seemed promising. One morning, when checking her social media, Paula saw a message from Matt, an old friend from high school. They scheduled a call to catch up, during which Matt told Paula that he was a civil engineer and after a few years working for a construction company he decided to start his own construction/real estate company. Matt's business concentrated on building apartments for early retirees. He was after the 60-year-old population because they had the financial power to spend on luxury apartments. Matt said that if done correctly, this was a good business to be in as the market would be growing for the coming decades! Paula was paying attention to Matt's business description as she made progress on her student loans, she continued saving in her 401(k) and just finished paying for her car.

Paula's car was still in great condition, so she was not planning on getting a new one in the upcoming years. She felt that she could use the money she was paying for her car for an investment. Paula wondered if she should invest in Matt's company. Matt described that due to the nature of his business, Paula's investment amount would have to be at least 150% of her car payment. Matt was working on a new apartment development in South Florida and said that she would get her money back in about 36 months, with a nice extra amount as a return on her investment. Matt gave Paula a week to consider the offer.

A few days after her conversation with Matt, Paula received a call from her Uncle Joey. Her Uncle Joey retired a few years back from a logistics firm and invested a large portion of his savings in a ride sharing business. His business was comprised of 10 cars with 20 drivers, so the cars were constantly in use, resulting in Joey making a nice salary. Joey explained that due to the miles driven he would have to replace about 50% of his fleet. However, instead of simply replacing vehicles he realized that he could make his company more competitive by providing additional services such as adding luxury cars and SUVs. Therefore, he explored his investment possibilities and found a car dealer that could deliver different types of cars, from small cars to SUVs and from basic to luxury. In addition, the car dealer offered an excellent price for the fleet and could deliver as early as next month. Therefore, Uncle Joey was looking for investors, and Paula believed this was a great opportunity for her. Moreover, the investment amount required for this business was the same as the car payment, so she would not need to allocate more money for this venture.

Paula felt that both options were very attractive, and she trusted both investors, but she did not have the money to invest in both and was not sure which option she should select. She would need to determine which option could deliver a higher return and then perform an analysis to determine if the extra amount required for Matt's project would deliver better financial results.

Reading Activity: Responding to Paula's Investment Story

- What information should be considered for her decision?
- Should Paula select the project by only considering the value of the rate of return?
- Should Paula select the project by only considering the investment amount?
- Which is more relevant, rate of return or investment amount?

Learning Outcomes

After covering this chapter, the student should be able to do the following:

1. Understand the rate of return concept and how to calculate it
2. How to use linear interpolation to solve rate-of-return problems
3. Understand the concept of incremental analysis and how to apply it for selecting mutually exclusive projects

Introduction: Rate of Return and Incremental Analysis

In the first chapters the concepts of interest rate and rate of return were introduced. These concepts describe a rate of equivalence derived from the concept of the time value of money. One of the first interest rate examples described the change in value for an item due to inflation. A following example considers the change in value of money from borrowing money. In this chapter the perspective is modified as the interest rate is based on the perspective from the lender, not the borrower. So, the interest rate is labeled as a rate of return instead of an interest rate.

Rate of Return Calculation

To understand this concept let us consider the equation to calculate the future value from a present value.

$$F = P(1 + i)^n$$

In previous chapters we calculated the future value, the present value, and in the previous chapter the number of periods, n. Therefore, the only variable left to calculate using this equation is the interest rate. From the future value equation, we may determine the interest rate as

Equation 6.1: Interest rate calculation from future value, present value and number of periods

$$i = \sqrt[n]{F/P} - 1$$

On the other hand, consider the annuities equation:

Equation 3.2: Annuity from a future value

$$A = \frac{F * i}{[(1+i)^n - 1]}$$

As you can see in Equation 3.7 calculating the interest rate from this equation becomes more complex in comparison to the FV equation. Therefore, calculating the interest rate from a different equation that involves multiple factors may prove difficult. Engineering economy problems for complex flows often require factors to be multiplied together, making this exercise mathematically complex to determine.

Fortunately, it is simpler to determine the value of an interest rate by equating present worth (PW), future worth (FW), and equivalent uniform annual worth (EUAW) to zero. The following Example 6.1 provides a description of how to determine an interest rate using PW.

Example 6.1

Jenny wants to buy a new machine for her business. The machine has a cost of \$10,000 and will generate an income of \$1,423.78 for 10 periods.

Therefore,
$PV = \$10{,}000$

$A = \$1{,}423.78$

$i = ?$

Following the PW approach, we obtain the following equation:

$$PW = -\$10{,}000 + \$1423.78\ (P/A, i\%, 10) = 0$$

$$PW = \frac{\$10{,}000}{\$1423.78} = 7.0236$$

To find the interest rate we can obtain the value from the compound interest tables provided in the appendix and look for the interest rate that provides a value of 7.0236 for a factor (P/A) for 10 periods.

Interest Rate	7.00%

	single payment		Uniform payment series				Arithmetic Gradient	
n	F/P	P/F	A/F	A/P	F/A	P/A	A/G	P/G
1	1.0700	0.9346	1.0000	1.0700	1.0000	0.9346	0.0000	0.0000
2	1.1449	0.8734	0.4831	0.5531	2.0700	1.8080	0.4831	0.8734
3	1.2250	0.8163	0.3111	0.3811	3.2149	2.6243	0.9549	2.5060
4	1.3108	0.7629	0.2252	0.2952	4.4399	3.3872	1.4155	4.7947
5	1.4026	0.7130	0.1739	0.2439	5.7507	4.1002	1.8650	7.6467
6	1.5007	0.6663	0.1398	0.2098	7.1533	4.7665	2.3032	10.9784
7	1.6058	0.6227	0.1156	0.1856	8.6540	5.3893	2.7304	14.7149
8	1.7182	0.5820	0.0975	0.1675	10.2598	5.9713	3.1465	18.7889
9	1.8385	0.5439	0.0835	0.1535	11.9780	6.5152	3.5517	23.1404
10	1.9672	0.5083	0.0724	0.1424	13.8164	7.0236	3.9461	27.7156
11	2.1049	0.4751	0.0634	0.1334	15.7836	7.4987	4.3296	32.4665

Fig. 6.1. Compound interest rate for 7%.

Example 6.1 (Continued)

To read this compound interest rate table consider the factor required: $(^P/_A, 7\%, 10)$. Then consider the number of periods, 10, highlighted in a rectangle in Fig. 6.1. with a value of 7.0236. Therefore, we consider a rate of return for this investment of 7%.

In Example 6.1 we learned that we could determine the rate of return by making the present worth equal to zero and then find the interest rate from the compound interest rate tables.

Exercise

Determine the rate of return for an investment of \$25,000 for 24 months that provides an annual benefit of \$1,811.77. Hint: Use interest rate tables.

Answer: 5%

A Word on Linear Interpolation

The results from Example 6.1 and the previous exercise have exact values that can be easily identified from the compound interest rate tables. Unfortunately, this is not always the case and frequently the interest rate values we require are not included in the compound interest rate tables. For example, the compound interest rate tables may include the values for 6% and 7% but may not include the values between 6% and 7%.

In such situations a practical solution is to use linear interpolation. Linear interpolation considers the following equation.

Equation 6.3: Linear interpolation equation

$$y_2 = \frac{(X_2 - X_1)(Y_3 - Y_1)}{(X_3 - X_1)} + Y_1$$

where Y_1 and Y_3 represent interest rate values and X_1 and X_3 represent the factor values that correspond to Y_1 and Y_3 Y_2 is the unknown interest rate value and X_2 is the known value for the unknown interest rate.

Example 6.2 shows how to acquire information from the two compound interest rate tables and how to apply the linear interpolation formula to solve a present worth problem.

Example 6.2

Fernando invested \$10,000 in a business. He received \$1,200 every year for the next 20 years. Determine the yearly interest rate.

P = \$10,000

n = 20 years

A = \$1,200

P = A (P/A, i%, 20)

\$10,000 = \$1,200 (P/A, i%, 20)

(P/A, i%, 20) = \$10,000 / \$1,200 = 8.3333

In order to interpolate to find the value of I we need P/A factors for n=20 that are lower and higher than the target value of 8.3333. The closer the values the higher level of accuracy.

Interest Rate 1.00%

	single payment		Uniform payment series				Arithmetic Gradient	
n	F/P	P/F	A/F	A/P	F/A	P/A	A/G	P/G
19	1.2081	0.8277	0.0481	0.0581	20.8109	17.2260	8.7017	149.8950
20	1.2202	0.8195	0.0454	0.0554	22.0190	18.0456	9.1694	165.4664
21	1.2324	0.8114	0.0430	0.0530	23.2392	18.8570	9.6354	181.6950

Fig. 6.2. Compound interest rate table for 1%.

Using the interest rate tables for $\left({}^{P}/_{A}, 1\%, 20\right) = 18.0456$

Interest Rate 15.00%

	single payment		Uniform payment series				Arithmetic Gradient	
n	F/P	P/F	A/F	A/P	F/A	P/A	A/G	P/G
19	14.2318	0.0703	0.0113	0.1613	88.2118	6.1982	5.2307	32.4213
20	16.3665	0.0611	0.0098	0.1598	102.4436	6.2593	5.3651	33.5822
21	18.8215	0.0531	0.0084	0.1584	118.8101	6.3125	5.4883	34.6448

Fig. 6.3. Compound interest rate table for 0.5%.

Use the interest rate tables for $\left({}^{P}/_{A}, 15\%, 20\right) = 6.2593$

Therefore, using the interest rate values we have the following:

Y (interest rate values)	X $\left({}^{P}/_{A}\right)$ factor values
$Y_1 = 1\%$	$X_1 = 18.0456$
$Y_2 = ?$	$X_2 = 8.3333$
$Y_3 = 15\%$	$X_3 = 6.2593$

Example 6.2 (Continued)

$$y_2 = \frac{(X_2 - X_1)(Y_3 - Y_1)}{(X_3 - X_1)} + Y_1$$

$$y_2 = \frac{(8.333 - 18.0456)(0.15 - 0.01)}{(6.2593 - 18.0456)} + 0.01$$

$$y_2 = \frac{(-9.7126)(0.14)}{(-11.7863)} + 0.01$$

$$y_2 = 0.115368 + 0.01 = 0.125368 = 12.5368\%$$

The exact interpolation value is 10.32%, this value was determined by using the excel function rate. Note that interpolated values are more accurate as they are closer to the intersection point.

In Example 6.2, we learned how to perform linear interpolation to find interest rate values. When performing a similar calculation consider using extreme interest rate values to generate your solution. The following example considers this perspective.

Exercise

Recalculate Example 6.2 considering an annuity of $750.

Answer: EUAC = 4.217%

Rate of Return for More Than One Factor

There are situations in which we want to calculate the rate of return, but the cash flow and its equation includes more than one factor. For example, a present worth calculation for a new car could consider the annuities of the monthly payments and the future value from the salvage value from selling the car.

To overcome this situation, use two interest rates to calculate the PW or the EUAW. You may interpolate values when the PW or the EUAW have different signs, for example interest rates of 1% and 20%. If the PW or the EUAW have the same sign, it is required to expand the range between rates until different signs are achieved. This situation is described in the following example.

Example 6.3

Kendra opened a pizza kiosk. Kendra invested $50,000 in the kiosk and received $1,500 every month for the next three years. After this time Kendra decided to sell her business share for $25,000. Determine the yearly rate of return.

Example 6.3

Present cost = \$50,000

Salvage value = \$25,000

Interest rate i =?% per year

n = 3 years = 36 months

A = \$1,500

$$PW = -\$50{,}000 + \$1{,}500\left(P/_{A,}\, i\%, 36\right) + \$25{,}000\left(P/_{F,}\, i\%, 36\right) = 0$$

Therefore, the interpolation target value is 0, and we try different interest values that would deliver positive and negative PW values.

Try 1%

$$PW = -\$50{,}000 + \$1{,}500\left(P/_{A,}\, 1\%, 36\right) + \$25{,}000\left(P/_{F,}\, 1\%, 36\right)$$

$$PW = -\$50{,}000 + (\$1{,}500 * 30.1075) + (\$25{,}000 * 0.6989)$$

$$PW = -\$50{,}000 + (\$45{,}161.26) + (\$17{,}473.12)$$

$PW = -\$50{,}000 + \$62{,}634.38 = \$12{,}634.38$ {Note this PW value is positive}

Try 20%

$$PW = -\$50{,}000 + \$1{,}500\left(P/_{A,}\, 20\%, 36\right) + \$25{,}000\left(P/_{F,}\, 20\%, 36\right)$$

$$PW = -\$50{,}000 + (\$1{,}500 * 4.9929) + (\$25{,}000 * 0.0014)$$

$$PW = -\$50{,}000 + (\$7{,}489.42) + (\$35.27) = -\$50{,}000 + \$7{,}524.69$$

$PW = -\$42{,}475.31$ {Note this PW value is negative}

Therefore, using the interpolation equation we have the following:

Y (interest rate values)	X $(P/_A)$ factor values
$Y_1 = 1\%$	\$12,634.38
$Y_2 =?$	0
$Y_3 = 20\%$	-\$42,475.31

$$y_2 = \frac{(X_2 - X_1)(Y_3 - Y_1)}{(X_3 - X_1)} + Y_1$$

$$y_2 = \frac{(0 - 12{,}634.38)(20\% - 1\%)}{(-42{,}475.31 - 12{,}634.38)} + 1\%$$

Example 6.3 (Continued)

$$y_2 = \frac{(-12{,}634.38)(0.19)}{(-\$55{,}109.69)} + 0.01 = 0.04355 + 0.01$$

$$y_2 = 0.05355 = 5.355\%$$

It should be noted that the interpolated interest rate value is per period; since the periods are in months, we must calculate the effective interest rate for the year:

$$i_{eff} = (1+i)^m - 1$$

$$i_{eff} = (1 + 0.05355)^{12} - 1 = 1.870087 - 1 = 87\%\ per\ year$$

Exercise

Determine the rate of return for a \$10,000 investment that provided a return of \$3,500 each year for five years and a salvage value of \$5,000.

Answer: 16.81% per year

Rate of Return for Annuities and Gradients

As described in previous chapters there are situations where annuities increase or decrease in a constant pattern either by a constant amount called arithmetic gradients or by a constant rate called geometric gradients. Example 6.4 describes how to determine the rate of return for an annuity with gradients.

Example 6.4

Bob is a 25-year-old engineering graduate who estimates that he invested \$200,000 in his undergraduate education. He also believes that his degree will allow him to earn \$30,000 per year on top of what he could have earned without a college degree and that his salary will increase by \$1,500 per year. Determine the rate of return for his education investment. Use EUAW to solve this problem.

$P = -\$200{,}000$
$A = \$30{,}000$
$G = \$1{,}500$

$$EUAW = -\$200{,}000\left({}^{A}/_{P}, i\%, 25\right) + \$30{,}000 + \$1{,}500\left({}^{A}/_{G}, i\%, 25\right) = 0$$

Example 6.4 (Continued)

$$EUAW = -\$200{,}000\left(A/_{P,}\, i\%, 25\right) + \$30{,}000 + \$1{,}500\left(A/_{G,}\, i\%, 25\right) = 0$$

Try 1%

$$EUAW = -\$200{,}000\left(A/_{P,}\, 1\%, 25\right) + \$30{,}000 + \$1{,}500\left(A/_{G,}\, 1\%, 25\right) = 0$$

$$EUAW = -\$200{,}000(0.0454) + \$30{,}000 + \$1{,}500\,(11.4831)$$

$$EUAW = -\$9081.35 + \$30{,}000 + \$17224.67 = \$38{,}143.32$$

Try 20%

$$EUAW = -\$200{,}000\left(A/_{P,}\, 20\%, 25\right) + \$30{,}000 + \$1{,}500\left(A/_{G,}\, 20\%, 25\right) = 0$$

$$EUAW = -\$200{,}000(0.2021) + \$30{,}000 + \$1{,}500\,(4.7352)$$

$$EUAW = -\$40{,}423.75 + \$30{,}000 + \$7{,}102.73 = -\$3{,}321.01$$

Y (interest rate values)	X $\left(P/_{A}\right)$ factor values
$Y_1 = 1\%$	\$38,143.32
$Y_2 = ?$	0
$Y_3 = 20\%$	-\$3,321.01

$$y_2 = \frac{(0 - 38{,}143.32)(0.2 - .01)}{(-\$3{,}321.01 - \$38{,}143.32)} + 0.01 = 18.4782\%$$

The rate of return after interpolating with MS Excel is 18.51%. Note that there is no effective interest calculation because the time periods are in years.

Exercise

Recalculate Example 6.4 considering an initial investment of \$150,000. Hint: select interpolation values between 20% and 25%

Answer: 23.95%

Rate of Return for Intermittent Periods

Previous examples show circumstances where every period has a cash flow. However, there are situations where cash flows are intermittent, meaning that not all periods have a cash flow. This is particularly applied in loans with a grace period; therefore, there is a lapse between receiving a loan and making the first payment. This is described in Example 6.5.

Example 6.5

Consider the case of Zach; he borrowed $15,000 before starting his first year of college and continued borrowing $15,000 for next three years. His loan agreement stipulated that no interest is charged during the first five years. Moreover, the loan includes a grace period of two years so his first payment will be made two years after graduation (year 6). In addition, Zach will have 10 years to pay his loan (until year 15) with a yearly interest rate of 5%. For simplicity we consider only one payment per year. Determine the rate of return for the lender.

This is a complex problem that requires multiple steps. First, to understand the problem, it is helpful to draw the cashflow.

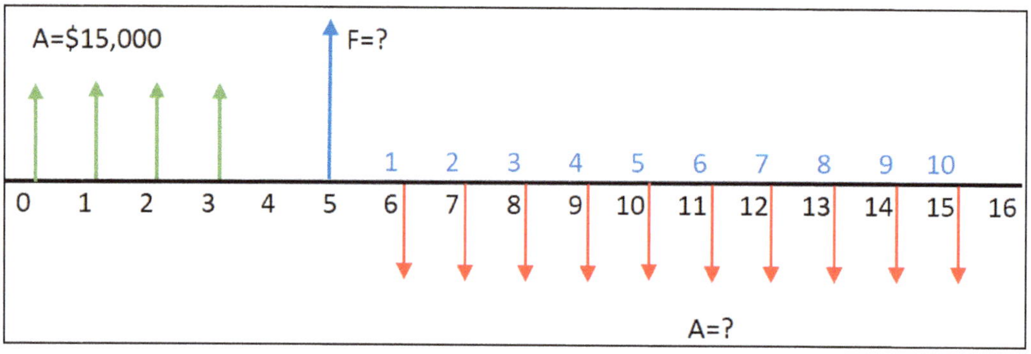

Fig. 6.4. Student loan cash flow problem Example 6.5.

The cash flow describes the following. Zach obtained the first loan in year 0 to pay for tuition in year 1 because tuition is paid at the beginning of the school year. Similarly, he had to borrow money at the beginning of the next three years.

The period numbers in black describe the periods from time 0 to the last payment in year 15. The periods in blue are used to describe the 10 payment periods. The amount of payment A is still to be determined.

Second, once we understand the cash flow, we must determine how much Zach owes in year 5. Remember payment annuities start in year 6, one period after the present value of year 5. Therefore, the loan has a value of $60,000 in year 5 because no interest is charged.

Third, since the debt for year 5 is $60,000, the next step to solving this problem is to calculate the yearly payments for the next 10 years:

$$Annuity = \$60{,}000 * \left({}^{A}/_{P}, 5\%, 10\right)$$

$$Annuity = \$60{,}000 * (0.1295) = \$7{,}770.00$$

Example 6.5 (Continued)

Fourth, we consider the cash flow from year 0 to year 15 to determine the rate of return.

Therefore, the equation is as follows:

$$PW = \$15{,}000 + \$15{,}000\left({}^{P}/_{A,}\, i\%, 3\right) - [\$7{,}700\ \left({}^{P}/_{A,}\, i\%, 10\right) * \left({}^{P}/_{F,}\, i\%, 5\right)]$$

When solving this equation for an interest rate of $i = 1\%$, the PW is

$$PW = \$15{,}000 + \$15{,}000\left({}^{P}/_{A,}\, 1\%, 3\right) - [\$7{,}700\ \left({}^{P}/_{A,}\, 1\%, 10\right) * \left({}^{P}/_{F,}\, 1\%, 5\right)]$$

$$PW = \$15{,}000 + \$15{,}000(2.941) - [\$7{,}700\ (9.471) * (0.9515)]$$

The PW for 1% is:

$$PW = \$\ \text{-}10{,}274.75$$

When solving this equation for an interest rate of $i = 10\%$, the PW is

$$PW = \$15{,}000 + \$15{,}000\left({}^{P}/_{A,}\, 10\%, 3\right) - [\$7{,}700\ \left({}^{P}/_{A,}\, 10\%, 10\right) * \left({}^{P}/_{F,}\, 10\%, 5\right)]$$

$$PW = \$15{,}000 + \$15{,}000(2.487) - [\$7{,}700\ (6.145) * (0.6209)]$$

The PW for 10% is:

$$PW = \$22{,}926.19$$

After performing an interpolation from the values obtained the interest rate is $i = 3.79\%$.

solving with Excel using the IRR function the interest rate is 2.95%.

What we learned from Example 6.5 is the following:

1. Although the problem statement describes a yearly interest rate of 5% for the loan, the rate of return for years 0 to year 15 was different.
2. To determine the internal rate of return all cash flows must be considered.

Exercise

Determine the internal rate of return for a car company that sells a car for $22,000 and allows customers to start making payments six months after they buy the car at $700 a month for 36 months, such that the customer will make the last payment in month 42. Hint: Calculate monthly and effective yearly interest rates.

Answer:
Monthly = 0.886%; Yearly = 11.1681%

Rate of Return for Two Buying Alternatives

In our daily life we encounter multiple situations in which we are offered two ways to buy a product or as in the initial reading, two investment alternatives.

For example, some companies allow customers to choose how they want to pay for an item. This serves a marketing purpose as customers are given a more affordable option. Paying "small" installments makes items more accessible, and it is also a way companies make additional income while showing that they want to support their customers.

Example 6.6

Ann is getting a streaming service. The streaming service provides the following two purchase options: \$119 per year or \$12.99 per month. Determine the internal rate of return for such service.

To solve this problem, it is required to determine the equation for each option and then equate them to find the IRR.

Option 1:

$$PW = \$119$$

Option 2:

$$PW = \$12.99 + \$12.99\,(P/A, i\%, 11)$$

Therefore, equating both Options

$$\$119 = \$12.99 + \$12.99\,(P/A, i\%, 11)$$

$$\frac{\$119 - \$12.99}{\$12.99} = 8.16089 = (P/A, i\%, 11)$$

We may then interpolate, and the interest rate for each period is 5.34%.

The interest rate per year is then calculated using the effective interest rate.

$$i_{eff} = (1 + 0.0534)^{12} - 1$$

$$i_{eff} = 1.866479 - 1 = 0.866479 = 86.6479\%$$

Example 6.6 shows that equating both alternatives allows us to find the IRR for the payments option. It should be noted that the first payment for the financing option is also done in month 0. In addition, a common error is to neglect the value of compound interest by multiplying the monthly cost by the number of periods. We should also remember to determine the yearly effective rate of return when the internal rate of return is calculated for periods smaller than one year.

Exercise

Determine the internal rate of return for a streaming service that has a monthly cost of $8 per month or a yearly cost of $80.

Answer: Monthly = 3.5%; yearly = 51.16%

In contrast, Example 6.7 shows the calculation of the IRR for a real estate investment for a future value that is higher than the present value.

Example 6.7

José made an investment three years ago. He bought a piece of land that had a price of $1,700,000. José has three questions. First, what is the IRR for his investment if the land has a current price of $2,300,000?

$$F = P(1+i)^n$$

$$i = \sqrt[3]{\frac{\$2{,}300{,}000}{\$1{,}700{,}000}} - 1 = 1.1060 - 1 = 10.60\% \text{ interest rate per period}$$

Second, José did not have enough money to pay the lump price in time 0, but, since this was a new development, he got an offer from the seller to make a down payment of $50,000 (in time 0), and then pay the reminder in 12 months (starting in month 1) at a 0% interest. Determine the yearly IRR for his investment if the land has a current price of $2,300,000.

Option 1: Obtaining the PW (year 0) from the current value in period (year 3)

$$PW = \$2{,}300{,}000\left({}^{F}/_{P,}\, i\%, 36\right)$$

Option 2: Obtaining the PW (year 0) for his investment pay $50,000 in time 0 and $137,500 for the following 12 months.

$$PW = \left[\$50{,}000 + \$137{,}500\left({}^{P}/_{A,}\, i\%, 12\right)\right]$$

Equating both options:

$$\$2{,}300{,}000\,\left({}^{P}/_{F,}\, i\%, 36\right) - \$50{,}000 - \$137{,}500\left({}^{P}/_{A,}\, i\%, 12\right) = 0$$

Try 1%:

$$\$2{,}300{,}000\,\left({}^{P}/_{F,}\, 1\%, 36\right) - \$50{,}000 - \$137{,}500\left({}^{P}/_{A,}\, 1\%, 12\right) = 0$$

$$\$2{,}300{,}000\,(0.6989) - \$50{,}000 - \$137{,}500(11.255) = \$9{,}907.50$$

Try 10%:

$$\$2{,}300{,}000\,\left({}^{P}/_{F,}\, 10\%, 36\right) + \$50{,}000 + \$137{,}500\left({}^{P}/_{A,}\, 10\%, 12\right) = 0$$

$$\$2{,}300{,}000\,(0.0323) + \$50{,}000 + \$137{,}500(6.8137) = -\$912{,}521.88$$

Example 6.7 (Continued)

After interpolating, the interest rate per month is $i = 1.10\%$. Therefore, the rate of return per year is:

$$i_{eff} = (1 + 0.011)^{12} - 1 = 1.14028 - 1 = 14.028\%$$

Third, if José had the opportunity to invest in the stock market, should José have invested in the Dow Jones Industrial Average (DJI), which had a return of 43% for the past three years? In other words, this is a yearly rate of 14.33%. Or should he have invested in the piece of land? Describe the advantages and disadvantages of his investment choice.

From Example 6.7 we found that the first question described a 10.60% rate of return per year. However, the second question shows that the rate of return is 14.02%. The difference between these two rates of returns is due to the monthly capitalization. Therefore, it is important to consider when payments are made.

Rate of Return For Project Selection (Incremental Analysis)

Example 6.7 describes that if José had the opportunity to invest in the Dow Jones or the piece of land, he should have invested in the Dow Jones. It seems obvious that when selecting between two projects the investor should always invest in the alternative with a higher rate of return. Unfortunately, following the highest rate of return as the only parameter to select an investment option may not be completely accurate.

The incremental analysis is an additional factor that allows identification of which option should be selected beyond the calculation of the internal rate of return (IRR). To describe what an incremental analysis is we first need to describe the concept of the minimum attractive rate of return (MARR). The MARR is a rate set by the investor and could be described as the lowest rate of return that the investor is willing to receive from an investment.

The incremental analysis for two alternatives requires the following steps:

1. Determine the IRR for each alternative.
2. Compare the IRR from each alternative to the MARR. If the IRR for an investment is lower than the MARR, this option is rejected. If both alternatives have IRR values higher than the MARR, then an incremental analysis is performed.
3. The incremental analysis calculates the difference between the two options and is calculated by subtracting values of the lowest-cost option from the values of the highest-cost option.
4. If the IRR from the incremental difference is higher than the MARR, then the alternative with the highest cost is selected; otherwise, the lowest-cost option should be selected.

Example 6.8 shows a comparison between two investing options.

Example 6.8

Michelle has two, one-year investment options. Michelle is looking for a rate of return of at least 6% per year. Therefore, Michelle's MARR is 6%.

- Option 1 requires an investment of $10,000 with a guaranteed return of $2,500.
- Option 2 requires an investment of $100,000 with a guaranteed return of $10,000.

If Michelle can only select one of the two investments, which option should she select? The information is provided in the Table 6.1.

Table 6.1. Incremental Analysis for Example 6.8

Option	Option 2	Option 1	Incremental difference Option 2 – Option1
Investment	$100,000	$10,000	$100,000 – $10,000 = $90,000
Return after 1 year	$10,000	$2,500	$10,000 – $2,500 = $7,500
IRR for alternative	$\frac{\$10,000}{\$100,000} = 10\%$	$\frac{\$2,500}{\$10,000} = 25\%$	$\frac{\$7,500}{\$90,000} = 8.33\%$

Since both options have a duration of 1 year, the IRR is obtained by dividing the return by the investment. For factors using longer periods or more complex flows, we use compound interest rate factors.

Once the IRR is calculated, since both IRRs are greater than the MARR of 6%, we require to perform an incremental analysis.

The IRR for the incremental analysis is 8.3%, which is higher than the 6% MARR. Based on these results, the alternative with the highest cost alternative (option 2) is selected even when its IRR is lower than the IRR from the other alternative.

From Example 6.8 we learned the following:

1. The alternative with the highest IRR does not necessarily mean that this alternative should be selected.
2. The MARR is set by the investor and should be compared with the IRRs for different alternatives.
3. When the IRR for the incremental difference is greater than the MARR the highest cost option should be selected.
4. The incremental analysis may justify the investment with the higher cost.

Exercise

Consider the information provided in Paula's story and determine which alternative Paula should select. Paula is an aggressive investor, and she will only consider projects that have an IRR of 15% per year.

Assume that Matt asks for an investment of $35,000; Matt states that the return on his project will be $40,500 after a period of three years, On the other hand, Joey promises a return of $28,000 for an investment of $19,500 after a period of three years. Hint: Divide the IRR value by 3 to have the yearly interest rate for one year.

Answer: Matt 28.80% | Joey 23.21% | Matt's investment, 41.33%, is higher than the MARR, 15%

Pairwise Comparison

From Example 6.8 we learned how to choose between two alternatives. However, it is likely that a project will have more than two alternatives. Therefore, the following methodology describes how to perform a pairwise comparison. This is shown in Example 6.9.

1. In a table, order the different alternatives from lowest to highest cost from left to right, such that the highest cost is in the last column.
2. Calculate the IRR for each alternative.
3. Compare the MARR with the IRR. Remove the alternatives that have an IRR below the MARR.
4. Perform a pairwise incremental analysis for the first pair of remaining alternatives. As before, if the IRR for the incremental is above the MARR, select the alternative with the highest cost; otherwise, select the alternative with the lowest cost.
5. Continue with the pairwise comparison until only one option remains.

Example 6.9

A business wants to buy new equipment to increase their capacity. They have the following options, all of them have a life of 10 years.

Option	Option A	Option B	Option C	Option D
Cost	$ 8,000.00	$ 6,500.00	$ 7,500.00	$ 9,500.00
Uniform annual benefit	$1,246.56	$968.69	$1,220.59	$1,290.75

Determine which option should be selected by finding the IRR for each alternative and considering a MARR of 7%.

Determine the IRR for each alternative, note the IRR the alternatives in this example do not require interpolation. However, this is not always the case.

Example 6.9 (Continued)

$$IRR_A \ \$8{,}000 = \$1{,}246.56\ (P/A, i\%, 10)$$

$$IRR_A = 6.4177 = (P/A, i\%, 10)$$

$$IRR_A = 9\%$$

$$IRR_B \ \$6{,}500 = \$968.69\ (P/A, i\%, 10)$$

$$IRR_B = 6.7101 = (P/A, i\%, 10)$$

$$IRR_B = 8\%$$

$$IRR_C \ \$7{,}500 = \$1{,}220.59\ (P/A, i\%, 10)$$

$$IRR_C = 6.1446 = (P/A, i\%, 10)$$

$$IRR_C = 10\%$$

$$IRR_D \ \$9{,}500 = \$1{,}290.75\ (P/A, i\%, 10)$$

$$IRR_D = 7.3601 = (P/A, i\%, 10)$$

$$IRR_D = 6\%$$

Since the MARR is 7%, alternative D should be rejected, and we should continue the pairwise comparison with the remaining options.

First, we reorder the alternative table from lowest to highest.

Option	Option B	Option C	Option A	Option D
Cost	$6,500.00	$7,500.00	$8,000.00	$9,500.00
Uniform annual benefit	$968.69	$1,220.59	$1,246.56	$950.00

Option C–option B are as follows:

$$Cost\ difference = (\$7{,}500 - \$6{,}500) = \$1{,}500$$

$$\text{UAB difference} = (\$1{,}220.59 - \$968.69) = \$251.90$$

$$\frac{\$1{,}500}{\$251.90} = 3.9698 = (P/A, i\%, 10)$$

Example 6.9 (Continued)

After interpolating, the interest rate per period is 21.637%. Therefore, option C is better and option B is rejected. We now proceed to comparing option A–option C.

option A–option C are as follows:

$$Cost\ difference = (\$8{,}000 - \$7{,}500) = \$500$$

$$\text{UAB difference} = (\$1{,}246.59 - \$1{,}220.59) = \$25.97$$

$$\frac{\$500}{\$25.97} = 19.2528 = (\text{P/A}, \text{i\%}, 10)$$

Interest Rate: 0.25%

	single payment		Uniform payment series				Arithmetic Gradient		
n	F/P	P/F	A/F	A/P	F/A	P/A	A/G	P/G	n
10	1.0253	0.9753	0.0989	0.1014	10.1133	9.8639	4.4794	44.1842	10
11	1.0278	0.9729	0.0898	0.0923	11.1385	10.8368	4.9750	53.9133	11
12	1.0304	0.9705	0.0822	0.0847	12.1664	11.8073	5.4702	64.5886	12
13	1.0330	0.9681	0.0758	0.0783	13.1968	12.7753	5.9650	76.2053	13
14	1.0356	0.9656	0.0703	0.0728	14.2298	13.7410	6.4594	88.7587	14

Fig. 6.5. Compound interest rate tables for 7.96%.

The value of 19.2528 is less than the value of 9.8639, with 0.25% interest from the compound interest rate tables. Therefore, the incremental IRR is less than the required 7% MARR. In fact, if you use the rate function from Excel, the interest rate is −10.42%.

Therefore, the alternative A is rejected, and alternative C is selected.

Example 6.10

Joey, Paula's uncle, as described in the reading, is considering buying 10 cars for his car-sharing business. According to his research, upscale services may charge more than basic car services. Calculate the cost and benefit information described in the following table for a time horizon of four years for all options. Joey has a MARR of 10%.

	Basic	SUV	Luxury	VIP
Cost	$ 20,000.00	$ 27,000.00	$35,000.00	$ 60,000.00
Yearly benefit after expenses	$6,584.69	$8,151.86	$10,803.40	$21,015.92

1. Calculate the IRR for each option.

2. Perform incremental analysis to determine which option should be selected or if Joey should not invest.

To find the internal IRR for the basic option the equation is

$$PW = -\$20{,}000 + \$6{,}584.69\ (P/A, i\%, 4)$$

$$\$20{,}000\ /\ \$6{,}584.69 = (P/A, i\%, 4)$$

Perform similar equations for all other options.

	Basic	SUV	Luxury	VIP
Cost	$20,000.00	$27,000.00	$35,000.00	$60,000.00
Yearly benefit after expenses	$6,584.69	$8,151.86	$10,803.40	$21,015.92
IRR	12%	8%	9%	15%

Since the SUV and the Luxury options have a MARR lower than 10%, the only financially feasible options are VIP and Basic. The following table shows the incremental analysis.

	VIP	Basic	VIP – Basic
Cost	$60,000.00	$20,000.00	$40,000.00
Yearly Benefit After Expenses	$21,015.92	$6,584.69	$14,431.23
IRR	15%	12%	16.48%

Since the IRR for the incremental is higher than 10%, the VIP is the best option.

Alternative Selection without a MARR

There are situations where the MARR is not provided. In this situation we can proceed with the process described in Example 4.8.

Option	Cost	Service Life in Years	Savings Per Year	Salvage Value
A	$4,484	3	$1,800	0
B	$7,985	6	$2,700	$1,500

This approach requires the calculation of PW or EUAW for two alternatives using different interest rates, as shown in Fig. 6.6.

i	PWA	PWB
1%	$687	$1,069
2%	$628	$946
3%	$571	$830
4%	$516	$721
5%	$464	$618
6%	$414	$521
7%	$366	$430
8%	$321	$343
9%	$277	$261
10%	$234	$184
11%	$194	$111
12%	$155	$41
13%	$117	-$24
14%	$81	-$87
15%	$46	-$146
16%	$13	-$202
17%	-$20	-$255
18%	-$51	-$306
19%	-$81	-$354
20%	-$110	-$400

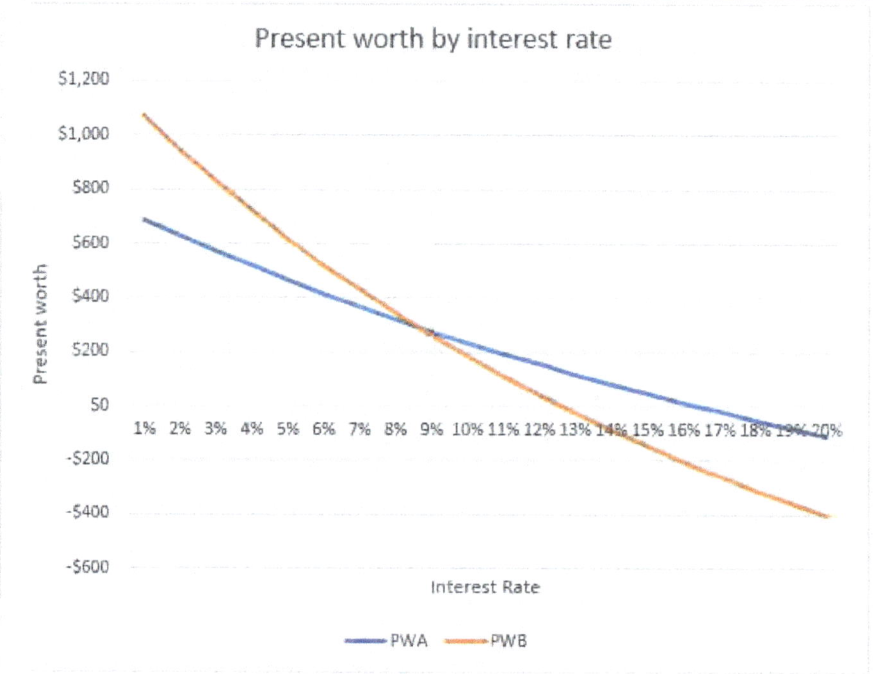

Fig. 6.6. PW from Example 4.8 for interest rates 1% to 20%.

Fig. 6.6. shows the PW changes for each interest rate and the graph depicts how the PW change. Once this information is obtained the next step is to generate a choice table, which option should be selected for a specific interest rate/MARR. Therefore, the choice table for Fig. 6.7. is

For a MARR of 1% to the intersection point 8.59%, select option B (the orange line) as it has a higher PW.

For a MARR of 8.59% to 16%, select option A.

For a MARR of more than 16%, no option should be selected as the PW of both alternatives is negative.

It is important to note that it is not possible to determine the exact intersectional value by simply looking at this figure. Instead, to find the intersection point, one must calculate the IRR from the PW equation for the incremental analysis from the following equation:

$$PW = -\$100 + \$1{,}050\left(P/F, i\%, 1\right) - \$50\left(P/A, i\%, 6\right) - \$1{,}300\left(P/F, i\%, 6\right)$$

The interest rate value is calculated using MS Excel as described in Table 6.1.

Table 6.1. Example 4.8 intersection point for the incremental analysis.

Period	Option A	Option B	Incremental
0	-\$1,900.00	-\$1,800.00	-\$100.00
1	\$1,400.00	\$300.00	\$1,100.00
2	\$250.00	\$300.00	-\$50.00
3	\$250.00	\$300.00	-\$50.00
4	\$250.00	\$300.00	-\$50.00
5	\$250.00	\$300.00	-\$50.00
6	\$250.00	\$1,500.00	-\$1,250.00

IRR = 8.59%

=IRR(W4:W10)

Example 6.11 presents a similar solution for four alternatives.

Example 6.11

Matt, Paula's friend, is considering four different vehicle options for his business. He plans on using this car to show his new apartment complex to potential customers. Each option has a different life, and he calculated the monthly benefts as the savings generated from avoiding having to rent a limo service. Matt does not have a MARR, and he wants to know which vehicle option he should purchase.

Table 6.3. Car options for Matt's car selection problem

Option	Vehicle cost, PV	Monthly annuity from savings, PMT	Salvage value, FV	Vehicle life, n
A	-\$15,000.00	\$ 1,150.00	\$ 500.00	48
B	-\$22,500.00	\$ 1,495.00	\$ 1,000.00	48
C	-\$30,000.00	\$ 1,633.00	\$ 3,000.00	48
D	-\$37,500.00	\$ 2,081.50	\$ 5,000.00	48

The solution for this example requires the calculation of the EUAW, which is calculated using the following equation:

$$EUAW = -PV\left(A/P, i\%, n\right) + Monthly\ annuity + FV\left(A/F, i\%, n\right)$$

Table 6.4 shows the EUAW values for the four vehicles using interest rates from 1% to 10%

Example 6.11 (Continued)

Table 6.4. EUAW values for four vehicle alternatives considering an interest rate from 1 to 10%

Interest rate	A	B	C	D
1%	$28,359.92	$33,650.81	$30,150.69	$38,441.49
2%	$20,080.82	$22,969.78	$18,929.59	$24,413.41
3%	$13,935.71	$15,031.73	$10,534.54	$13,882.66
4%	$9,298.30	$9,034.53	$4,155.06	$5,856.69
5%	$5,740.66	$4,429.21	($768.43)	($353.11)
6%	$2,967.03	$835.79	($4,626.50)	($5,229.46)
7%	$770.61	($2,011.81)	($7,694.74)	($9,114.35)
8%	($994.93)	($4,302.11)	($10,169.75)	($12,252.66)
9%	($2,434.38)	($6,170.28)	($12,193.41)	($14,821.65)
10%	($3,623.69)	($7,714.40)	($13,869.24)	($16,951.09)

The highlighted values show the highest EUAW for the different options considering each row's interest rate.

The choice table then is :

0% – 2.6% Option "D," yellow line

2.61% – 3.75% Option "B," green line

3.76% – 8% Option "A," blue line

The information provided in Table 6.4 is plotted in Fig. 6.8.

The intersection points are the range values described in the choice table. For example, the first range of values goes from 0% to 2.6%; for these interest rates option B is the choice that the delivers the highest EUAW of the available options.

In addition, option C is not considered at any range as it fails to deliver an EUAW value that is higher than the EUAW from the other alternatives.

Note that these values are not exact as they are taken from the Fig. 6.8. To determine the exact values, it is required to perform a pairwise incremental analysis between the different alternatives. This is shown in the Excel solution at the end of this chapter.

Example 6.11 (Continued)

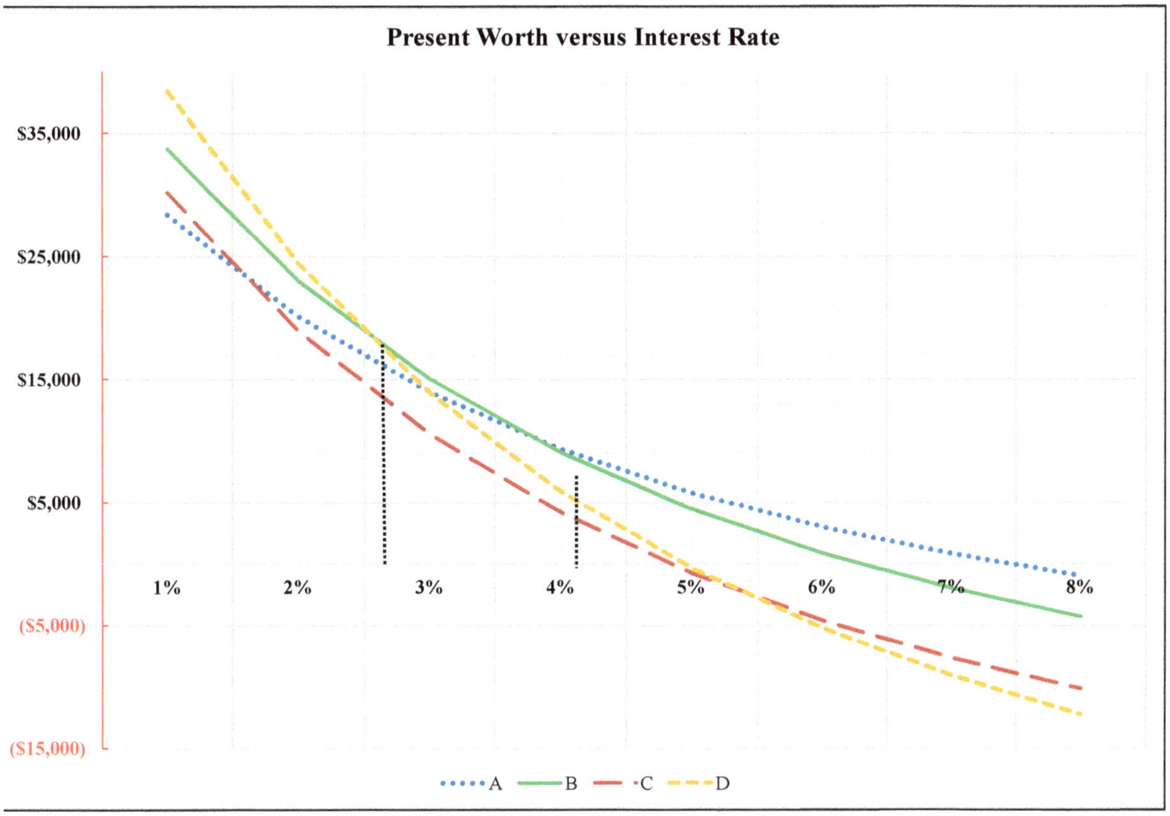

Fig. 6.7. EUAW for the different car options for Example 6.11.

Using MS Excel to Solve Examples

As in previous chapters, the following section shows how to solve the previous examples using MS Excel. Note that for IRR calculations, PV and annuity values must have opposite signs.

Example 6.1: Excel Solution

Determine the IRR for buying a new machine. Assume that the machine has a cost of $10,000 and will generate an income of $1,423.78 for 10 periods.

$PV = -\$10{,}000$

$A = \$1{,}423.78$

$i = ?$

Example 6.1: Excel Solution (Continued)

Determine the IRR for buying a new machine. Assume that the machine has a cost of $10,000 and it will generate an income of $1,423.78 for 10 periods.

To solve this equation, we use the Excel formula IRR and then select the values for PV, annuity, and number of periods.

	A	B	C	D
1				
2		PV	-$10,000.00	
3		Annuity	$1,423.78	
4		n	10	
5		interest rate	7.00%	=RATE(C4,C3,C2)
6				

Fig. 6.8. Excel solution Example 6.1.

This solution uses the MS Excel function RATE, which requires the present value, the annuity PMT, and the number of periods, n, to be entered.

Example 6.2: Excel Solution

Fernando invested $10,000 in a business. He received $1,200 every year for the following 20 years. Determine the yearly interest rate.

$PV = -\$10{,}000$

$A = \$1{,}200$

$n = 20\ years$

$i = ?$

To solve this equation, we use the Excel formula IRR and then select the values for PV, annuity, and number of periods. Since the data is in years there is no need to calculate an effective interest rate.

	A	B	C	D
1				
2		PV	-$10,000.00	
3		Annuity	$1,200.00	
4		n	20	
5		interest rate	10.32%	=RATE(C4,C3,C2)
6				

Fig. 6.9. Excel solution Example 6.2.

This solution uses the same method as that described in the previous example.

Example 6.3: Excel Solution

Kendra's friend opened a pizza kiosk. Kendra invested \$50,000 in the kiosk and received \$1,500 every month for the next three years. After this time Kendra decided to sell her business share for \$25,000. Determine the yearly rate of return.

$PV = -\$50,000$

$A = \$1,500$

$F = \$25,000$

$n = 36\ months$

$i = ?$

	A	B	C	D
1				
2		PV	-\$50,000.00	
3		Annuity	\$1,500.00	
4		Future	\$25,000.00	
5		n	36	
6		interest rate	2.0470%	=RATE(C5,C3,C2,C4)
7				
8		Effective interest rate	27.5276%	=EFFECT(12*0.79%,12)
9				

Fig. 6.10. Excel solution Example 6.3.

The solution for the interest rate is found by using the function RATE. Since this is the rate per month it requires the function EFFECT to calculate the effective rate of return per year. It should be noted that this calculation differs from the previous by-hand solution as the manual interpolation ranges were very large. Closer interpolation values will provide a better approximation.

Example 6.4: Excel Solution

Bob is a 25-year-old engineering graduate who estimates that he invested \$200,000 in his education. He also believes that his degree will allow him to earn \$30,000 per year on top of what he could have earned without a college education and that his salary will increase by \$1,500 per year. Determine the rate of return for his education investment. Consider 25 periods and solve using annuities.

To solve Example 6.4, we created a cash flow for every period, from period 0 to period 25.

The initial flow for year 0 is -\$200,000, and the cash flow for period 1 is \$30,000. For the remaining periods it increased by \$1,500.

Once the flows are completed, we enter the function IRR from Excel and select the range of cells required. This is the interest rate per period. In this case one period is one year.

	A	B	C	D
1				
2		Period	Cash flow	
3		0	-\$200,000.00	
4		1	\$30,000.00	
5		2	\$31,500.00	
6		3	\$33,000.00	
7		4	\$34,500.00	
8		5	\$36,000.00	
9		6	\$37,500.00	
10		7	\$39,000.00	
11		8	\$40,500.00	
12		9	\$42,000.00	
13		10	\$43,500.00	
14		11	\$45,000.00	
15		12	\$46,500.00	
16		13	\$48,000.00	
17		14	\$49,500.00	
18		15	\$51,000.00	
19		16	\$52,500.00	
20		17	\$54,000.00	
21		18	\$55,500.00	
22		19	\$57,000.00	
23		20	\$58,500.00	
24		21	\$60,000.00	
25		22	\$61,500.00	
26		23	\$63,000.00	
27		24	\$64,500.00	
28		25	\$66,000.00	
29				
30			18.51%	IRR(C3:C28)
31				

Fig. 6.11. Excel solution Example 6.4.

Example 6.5: Excel Solution

Zach, he borrowed $15,000 before starting his first year of college and he continued borrowing $15,000 for the next three years. Consider that his loan agreement stipulated that no interest is charged in the first five years.

Moreover, the loan includes a grace period of two years, so his first payment will be made two years after graduation (year 6). In addition, Zach will have 10 years to pay his loan (until year 15) with a yearly interest rate of 5%. For simplicity we consider only one payment per year. Determine the rate of return for the lender.

Column C shows the original borrowing amounts and the amounts paid for 10 periods starting in period 6.

Column D shows the borrowing amount in year 5.

Cell C:20 shows the calculated value from the function =IRR(C3:C18). The internal rate of return per period considers period 0 cell C:3 to period 15 cell C:18

C20 =IRR(C3:C18)

	A	B	C	D
1				
2	Period	Payment	Cash flow	
3	0		$15,000.00	
4	1		$15,000.00	
5	2		$15,000.00	
6	3		$15,000.00	
7	4		$0.00	
8	5		$0.00	$60,000.00
9	6	1	-$7,770.00	
10	7	2	-$7,770.00	
11	8	3	-$7,770.00	
12	9	4	-$7,770.00	
13	10	5	-$7,770.00	
14	11	6	-$7,770.00	
15	12	7	-$7,770.00	
16	13	8	-$7,770.00	
17	14	9	-$7,770.00	
18	15	10	-$7,770.00	
19				
20			2.95%	

Sum of amount borrowed in

10 equal payments interest rate 5% per

Fig. 6.12. Excel solution Example 6.5.

Example 6.6: Excel Solution

Ann is getting a streaming service. The streaming service provides the following two purchase options: $119 per year or $12.99 per month. Determine the internal rate of return for such service.

The solution requires the net value for period 0 to be calculated, which is the difference between $119 and $12.99. This tracks as both payments are required in period 0.

The following periods, 1–11, have negative flow values of $12.99.

As before, the internal rate of return is calculated using the function IRR for the monthly rate of return.

The function EFFECT requires the nominal interest rate, which is obtained by multiplying the rate of 5.34% by 12 periods and the number of periods per year.

These functions are described below the calculated values.

C16 =IRR(C3:C14)

	A	B	C	D
1				
2		Period	Cash flow	
3		0	$106.01	
4		1	-$12.99	
5		2	-$12.99	
6		3	-$12.99	
7		4	-$12.99	
8		5	-$12.99	
9		6	-$12.99	
10		7	-$12.99	
11		8	-$12.99	
12		9	-$12.99	
13		10	-$12.99	
14		11	-$12.99	
15				
16			5.34%	
17			IRR(C3:C14)	
18			86.648%	
19			=EFFECT(D19*12,12)	

10 equal payments interest

Fig. 6.13. Excel Solution Example 6.6.

Example 6.7: Excel Solution

José made an investment three years ago. He bought a piece of land that had a price of $1,700,000. José has three questions. First, what is the IRR for his investment if the land has a current price of $2,300,000?

	A	B	C	D	E
1					
2		PV	-$1,700,000.00		
3		FV	$2,300,000.00		
4		n	3		
5		interest rate	10.60%	=RATE(C4,C3,C2)	
6					

Use function RATE to calculate the interest rate per period.

Fig. 6.14. Excel Solution Example 6.7.

Second, José did not have enough money to pay the lump price in time 0, but, since this was a new development, he got an offer from the seller to make a down payment of $50,000 (in time 0), and then pay the reminder in 12 months (starting in month 1) at 0% interest. Determine the yearly IRR for his investment if the land has a current price of $2,300,000.

	A	B	C
1			
2		Period	Flow
3		0	$50,000.00
4		1	$137,500.00
5		2	$137,500.00
6		3	$137,500.00
7		4	$137,500.00
8		5	$137,500.00
9		6	$137,500.00
10		7	$137,500.00
11		8	$137,500.00
12		9	$137,500.00
13		10	$137,500.00
14		11	$137,500.00
15		12	$137,500.00
16		13	$ -
17		14	$ -
37		34	$ -
38		35	$ -
39		36	-$2,300,000.00
40			
41			1.021%
42			=IRR(C9:C45)
43			
44			0.129645207
45			=EFFECT(C47*12,12)
46			

The solution requires the payment options and then the value of the property after 36 months.

The function IRR is used to find the internal rate of return, and the function EFFECT is used to calculate the effective yearly rate.

To show the solution, rows 15 to 33 are hidden and have a value of $0.00.

Fig. 6.15. Excel Solution Example 6.7.

Example 6.9: Excel Solution

A business wants to buy new equipment to increase their capacity. They have the following options, all of them with a useful life of 10 years.

Option	Option A	Option B	Option C	Option D
Cost	$ 8,000.00	$ 6,500.00	$ 6,500.00	$ 9,500.00
Uniform annual benefit	$1,594.02	$1,150.40	$1,446.35	$1,892.89

Determine the IRR for each alternative and assuming a MARR of 13%, determine which option should be selected.

	A	B	C	D	E	F
1						
2		Option Cost	Option A	Option B	Option C	Option D
3			$8,000.00	$6,500.00	$6,500.00	$9,500.00
4		Uniform annual benefit	$1,594.02	$1,150.40	$1,446.35	$1,892.89
5		IRR	15.00%	12.00%	18.00%	15.00%
6			=RATE(10,-C4,C3)	=RATE(10,-D4,D3)	=RATE(10,-E4,E3)	=RATE(10,-F4,F3)
7						
8			Incremental analysis A - C			
9		Option Cost	Option A	Option C		
10			$8,000.00	$6,500.00	$1,500.00	
11		Uniform annual benefit	$1,594.02	$1,446.35	$147.67	
12		IRR	15.00%	18.00%	-0.28%	
13						
14			Incremental analysis D - C			
15		Option Cost	Option D	Option C		
16			$9,500.00	$6,500.00	$3,000.00	
17		Uniform annual benefit	$1,892.89	$1,446.35	$446.54	
18		IRR	15.00%	18.00%	7.97%	
19						

Fig. 6.16. Excel solution for Example 6.9.

The solution for Example 6.10 follows a similar procedure.

Example 6.11: Excel Solution

Matt, Paula's friend, is considering four different vehicle options for his business. He plans on using this car to show his new apartment complex to potential customers. Each option has a different life, and he calculated the monthly benefts as the savings generated from avoiding having to rent a limo service. Matt does not have a MARR, and he wants to know which vehicle option he should purchase.

Table 6.3: Car options for Matt's car selection problem

Option	Vehicle Cost	Monthly annuity from savings PMT	Salvage value FV	Vehicle life n
A	-$15,000.00	$ 1,150.00	$ 500.00	48
B	-$22,500.00	$ 1,495.00	$ 1,000.00	48
C	-$30,000.00	$ 1,633.00	$ 3,000.00	48
D	-$37,500.00	$ 2,081.50	$ 5,000.00	48

To solve this problem, we need to use the solver function in Excel. First it is required to re rearrange the data to use the interest rate from the cell and then to calculate the PW for each alternative.

H3 f_x =D3+PV(C3,G3,-E3,F3)

	A	B	C	D	E	F	G	H
1								
2			interest	Alternative Cost (PV)	PMT	FV	n	PW calculation
3		A	5%	-$15,000.00	$ 1,150.00	$ 500.00	48	$5,740.66
4		B	5%	-$22,500.00	$ 1,495.00	$ 1,000.00	48	$4,429.21
5		C	5%	-$30,000.00	$ 1,633.00	$ 3,000.00	48	($768.43)
6		D	5%	-$37,500.00	$ 2,081.50	$ 5,000.00	48	($353.11)
7			5%					
8							D-B	($4,782.32)

Fig. 6.17. PV calculation following the same interest rate.

Second, the interest rate for all alternatives should follow the value set in one alternative as described in Fig. 6.14 the interest rate for each alternative follows the interest rate value from cell C7. This is done by making cell C3 = C7, cell C4 = C7, cell C5 =C7, and cell C6 =C7.

This is done, so when we change the value in C7 it changes the interest rate values for all alternatives.

Third, enter the PV formula in column H. This is done by entering the present value for each alternative in column D, and then adding the PV derived from the interest rate (column C), the annuity (column E), the salvage value (column F), and the number of periods (column G). It is important to note that since the cost is entered as a negative value the payment (PMT) should be entered as a negative. Remember that Excel automatically changes the signs.

Fourth, once the PW is calculated for each alternative, set a cell that will subtract the alternatives that we know have an intersection.

Example 6.11: Excel Solution (Continued)

Fifth, from the information provided in Fig. 6.8 we know that there is an intersection with option D and option B. Therefore, cell H8 calculates the difference between H6–H4 with the formula H8 = H6 – H4.

Sixth, once the table has been set, we select cell H8, and with this cell selected click the "Data" option from the menu. Once the "Data" option has been selected, select "What if Analysis"; from there select "Goal Seek."

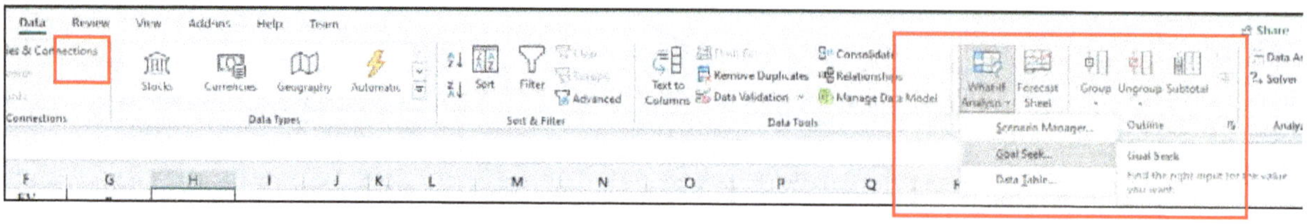

Fig. 6.18. Data menu and What If Analysis menu.

Seventh, selecting the option "Goal Seek" opens a new window, as shown.

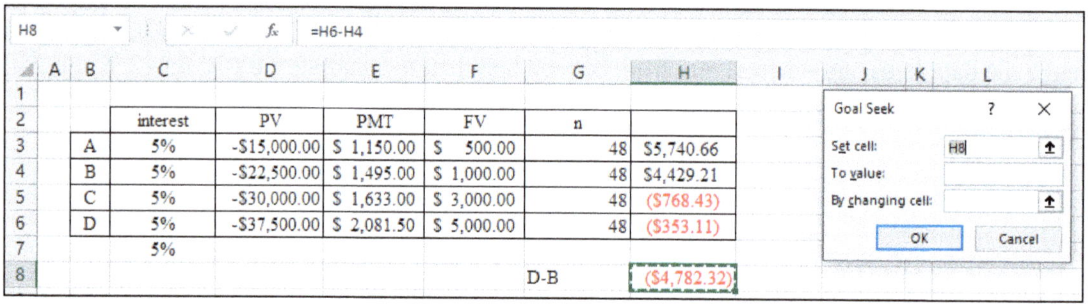

Fig. 6.19. Goal Seek menu.

Then we enter the following. To value: 0. This is because when both functions intersect there is no difference between them. Finally, we select the cell that will be changing in this case the interest rate, so we enter C7 for the option by changing cell as shown in Fig. 6.21

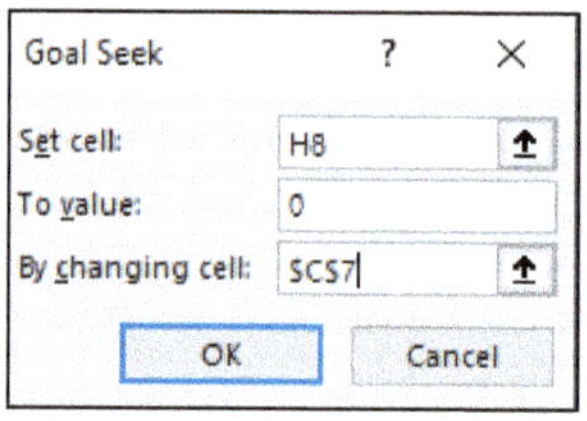

Fig. 6.20. Goal seek menu with correct parameters.

These parameters tell Excel to perform multiple iterations in the interest rate until it finds no difference between the present worth values from option D and option B.

After you click OK, Excel shows different iteration values until it finds the interest rate that shows an intersection between PW for the different options; it is 2.53% for options D and B. If you follow similar steps, you will find an intersection between A and B at an interest rate of 3.78%.

What Did We Learn in This Chapter?

1. We learned the relevance of calculating an internal rate of return.
2. The internal rate of return is a way to evaluate the viability of a project, even if the cash flow of the project includes gradients.
3. A minimum attractive rate of return (MARR) allows us to select an investment project based on a rate of return that will deliver a minimum desired return.
4. When two or more alternatives have an internal rate of return equal to or greater than the MARR, performing an incremental analysis allows us to determine which alternative should be selected.
5. When analyzing two or more projects without a MARR we can apply different interest rates to which alternative should be selected.
6. Excel provides a useful tool to quickly calculate rates of returns for a cash flow.

Review Questions

Directions: Refer to what you learned in this chapter to answer the following questions:

1. What is the difference between an interest rate and an internal rate of return?
2. How would you define and determine the MARR for a project?
3. Can a project be evaluated without a MARR?
4. Given the opportunity to pay for a streaming service monthly or yearly, which payment alternative should be selected?
5. What is incremental analysis, and why is it useful?
6. Should all investment projects be considered an IRR?
7. Is it possible to determine an IRR for an expense project, such as buying a house?

Chapter 6 Practice Problems

Use the following problems to practice your newly acquired knowledge. Answers to even problems are in the back of the book.

1. A new machine can be bought for $1,000,000. It is estimated that this machine will generate savings of $125,000 per year and that the machine does not have a salvage value at the end of 10 years. Determine the internal rate of return for this equipment.

2. Manuel, a graduate student, estimates that his master's degree cost $100,000. He believes that he will be able to earn $8,000 more than a person without a master's degree. Determine the internal rate of return for his master's education assuming he plans to work for the next 30 years.

3. Solve problem 2 using excel and assuming that Manuel will be able to get a salary increase of 2% each year.

4. Nichole, a recent engineering graduate, works in a car seat manufacturing firm. Each seat requires a metal frame support. Nichole's job is to analyze if the company should buy a new machine to improve the production rate. The new machine will generate savings of $0.2 per seat with a monthly production of 50,000 seat units per month. The new machine has a cost of $150,000 and can be sold after 15 years of use for $50,000. Nichole's company uses a MARR of 5% per year.

5. A college in the Midwest is thinking about opening a new sports field. Since their football team is very popular, they calculate that they could get, on average, 20,000 fans every game each semester. Each semester the team will play eight local games. Ticket price, on average, is $50 per ticket. Consider that the college can keep 70% of the ticket profits and will need to use the other 30% for team expenses. Determine the rate of return for this project if the cost of the arena is $75,000,000.

6. Hector is considering two types of compressors for his company compressor A has a cost of $100,000 and has operation and maintenance savings of $13,000 per year. Compressor B has a cost of $80,000 with yearly operation and maintenance savings of $15,500. Assume a time horizon of 10 years for both alternatives and a MARR of more than 15%. Determine the following:

    a. The IRR for each alternative

    b. The IRR for each alternative considering a salvage value of 15% of their original cost.

    c. Which if any alternative should be selected. If needed perform an incremental analysis.

7. Determine the IRR from selling a car. The car has a price of $23,500, the car dealer is charging an interest rate of 0.25% per month and the buyer is required to make the first payment five months after buying the car.

8. A company invested last year in a new machine with a cost of $2000. The first three months they had a cash flow of $0, the next three months they had a flow of $300, the next three months they had a flow of $150, and the last three months they had a flow of $300. Did they achieve their MARR of 20%?

9. Calculate the IRR for the following flow:

Year	0	1	2	3	4	5	6	7	8
Amount	$20,000	$0	$0	−$10,000	−$7,500	−$5,000	−$2,500	$0	$2,500

10. Calculate the IRR for the following flow:

Year	0	1	2	3	4	5	6	7	8
Amount	$30,000	−$6,000	−$4,000	−$6,000	−$4,000	−$6,000	−$4,000	−$6,000	−$4,000

11. A local company is analyzing five alternatives for a new machine; the information is given in the following table.

Option	Cost	Savings per Year	Salvage Value, FV	Vehicle Life, n
A	−$10,000.00	$ 2,050.00	$ 500.00	5
B	−$12,500.00	$ 3,405.00	$ 1,000.00	5
C	−$15,300.00	$ 3,680.00	$ 3,000.00	5
D	−$20,500.00	$ 5,281.50	$ 5,000.00	5
E	−$30,125.00	$ 6,785.00	$ 5,000.00	5

a) Which alternative should be selected without considering salvage value? The company is considering a MARR of 7%.

b) Which alternative should be selected with considering salvage value? The company is considering a MARR of 10%.

Chapter 7 Engineering Economy Metrics and Uncertainty

Pepita's Food Truck Story

IMG 7.1

Diego, Paula's brother, has invested in the food truck business for the past 10 years in LA. He acquired a food truck and named it Pepita's Food. His business plan was simple: using his contacts in the entertainment business he provided catering services for movie and TV productions.

Diego's business was doing well, but during the COVID-19 health crisis his business was severely hit. Some productions got cancelled, and others switched locations; to make matters worse, he had to downsize his business, and now that things have started to come back to "normal" he faces a new problem as: it is difficult to find staff. Therefore, Diego is looking for new areas where he can expand his operation. His objective is clear: find an area with a lower level of uncertainty even if it reduces his profit margin. Caleb, Diego's friend, is a superintendent for a large construction company. Caleb told Diego that his company is highly likely to get a big infrastructure contract to renovate a sewer system. This renovation project will require the construction team to move to different remote locations and to work at odd hours, so the construction company will need on-site food services.

At first this expansion plan seemed reasonable to Diego; he knew he could increase his income. However, there were a few things that made Diego worry. The first thing was the length of the project. Caleb said that once they started the renovation, and they acquired more information about the sites, the length of the project could be modified, so the project could take between one to three years. The second concern was the size of the project it; could go from a "simple" restoration to a project that would require additional construction bringing uncertainty in the number of employees required. This was still unknown because the state government had to approve the funds required and even the approval the construction, company still needed to negotiate the contract with the union. Finally, the expansion would require an additional investment in one or two new trucks, which would also mean the additional expense of hiring new employees.

Diego is unsure about what to do. Should he continue to serve his current customer base? Should he move his attention to the construction company? Or should he try serving both? In such a case, how long would it take for him to recover his investment? Will he be able to increase his income from this investment? How should he address the certainty from his current operation with the uncertainty of the business expansion?

Reading Activity: Responding to Pepita's Food Truck Story

1. How long should it take for him to recover his investment?
2. How should Diego evaluate the different alternatives? Will he be able to increase his income?
3. How should he consider uncertainty in his analysis?
4. What is the role of historical information in his analysis?

Learning Outcomes

After covering this chapter, the student should be able to:

1. Learn the calculation of the number of periods to recover an investment.
2. Learn how to use benefit /cost analysis to select an investment for the private and public sector.
3. Learn how to generate estimates for uncertain data.
4. Learn how to consider uncertainty for the selection of an investment project.
5. Learn the role of probability and historical data in investment projects.

Introduction: New Comparison Methods and Uncertainty

Until this point, we have considered different methods to determine the best financial alternative. We learned how to calculate the present worth (PW), equivalent uniform annual worth (EUAW), and the rate of return (IRR) for a given project. However, there are some additional ways to evaluate a project. Furthermore, as described in the introduction, we live in a world full of uncertainty and that should be considered in a financial analysis.

The following sections introduce new concepts to determine the payback period, benefit-cost ratio (B/C), and future worth (FW) for engineering economy projects. These methods should not be used to substitute the previously described methods and are a way to supplement the previous methods. The later part of the chapter considers the incorporation of uncertainty and probability in engineering economy projects.

Payback Period

IMG 7.2

One common investment question, if not the first, is how long it takes to recover an investment. This is portrayed in the TV show *Shark Tank* which presents entrepreneurs who invite a group of five investors to participate in their business. Typically, an entrepreneur offers a share of their business for an investment amount that they will use to expand the business. Kevin O'Leary, one of the "sharks," often asks the question "When will I have my money back?"

O'Leary's approach is called the payback period. This analysis, although simplistic, gives an initial insight into the viability of an investment project. This approach is the only one in this textbook that determines the number of periods to recover an investment without considering an interest rate or a rate of return, nor the cash flow after the investment has been recovered, such as additional inflows or outflows of money, including a salvage value. Example 7.1 describes this situation.

Example 7.1

Determine the best alternative based on the payback period from the alternatives presented in Example 6.8 assuming that the return is constant for multiple years.

Table 7.1 Incremental Analysis Example 6.8

Option	Option 2	Option 1
Investment	$100,000.00	$10,000.00
Yearly return	$10,000.00	$2,500.00

The number of periods required to recover each investment is calculated as

Equation 7.1: Payback period

$$n = \frac{Cost}{Yearly\ return}$$

For option 1 it shows the following:

$$n = \frac{\$10{,}000}{\$2{,}500} = 4\ periods$$

And for option 2 it shows the following:

$$n = \frac{\$100{,}000}{\$10{,}000} = 10\ periods$$

Therefore, option 1 should be selected as it takes fewer periods to recover the initial investment. It should be noted that this contrasts with the answer we obtained in Example 6.8, where option 2 was selected because it was based on the internal rate of return and an incremental analysis.

Example 7.2 shows a case with uneven income and salvage values.

Example 7.2

Determine the best alternative based on the payback period for the following alternatives:

	Cost Year 0	Savings Year 1	Savings Year 2	Savings Year 3	Savings Year 4	Savings Year 5	Savings Year 6	Salvage Year 7
Alternative A	$10,000	$500	$1,000	$2,000	$4,000	$10,000	$15,000	$0
Alternative B	$20,000	$5,000	$5,000	$10,000	$10,000	$15,000	$15,000	$35,000

Example 7.2 (Continued)

Duration for alternative A:

	Cost Year 0	Savings Year 1	Savings Year 2	Savings Year 3	Savings Year 4	Savings Year 5	Savings Year 6	Salvage Year 7
Alternative A	$10,000.00	$500	$1,000	$2,000	$4,000	$10,000	$15,000	$0
Unrecovered amount		$9,500	$8,500	$6,500	$2,500	$2,500		
Recovery periods (n)		1	2	3	4	4.25		

Duration for alternative B:

	Cost Year 0	Savings Year 1	Savings Year 2	Savings Year 3	Savings Year 4	Savings Year 5	Savings Year 6	Salvage Year 7
Alternative A	$20,000.00	$5,000	$5,000	$10,000	$10,000	$15,000	$15,000	$35,000
Unrecovered amount		$15,000	$10,000	$0				
Recovery periods (n)		1	2	3				

Only one quarter of the savings from year 5 is required to recover the cost of alternative A. On the other hand, alternative B requires three full years to recover its cost.

Therefore, alternative B should be selected as it has a shorter recovery period. Furthermore, it should be noted that neither the additional savings nor the salvage values are considered.

Exercise

Calculate the payback period for the following alternatives:

	Cost	Savings Y1	Savings Y2	Savings Y3	Savings Y4	Savings Y5	Savings Y6
Alternative 1	$20,000.00	$3,500	$4,500	$5,500	$6,500	$7,500	$8,500
Alternative 2	$35,000.00	$8,000	$10,500	$13,000	$15,500	$18,000	$20,500

Answer Payback = Alt 1 (4 years) | Alt 2 (3.23 years)

From Example 7.1 and 7.2 we learned the following:

1. The payback period and discounted payback period calculate the number of periods required to recover an initial cost.
2. The calculation of the number of periods may and may not consider an interest rate.

Discounted Payback Period

In contrast with the payback, the period discounted payback period considers the interest rate to determine the number of periods. To solve with this approach, we write a PW or worth EUAW equation to determine the number of periods, as shown in Example 7.3.

Example 7.3

Redo Example 7.1 using payback period from the alternatives presented in Example 6.8 assuming that the return is constant for multiple years and has an interest rate of 5% per year.

Table 7.2 Data from Example 6.8

Option	Option 2	Option 1
Investment	$100,000.00	$10,000.00
Yearly return	$10,000.00	$2,500.00

Option 1:

$$PW = -\$10{,}000 + \$2{,}500\ \left({}^{P}/_{A}, 5\%, n\right)$$

$$\left({}^{P}/_{A}, 5\%, n\right) = \frac{\$10{,}000}{\$2{,}500} = 4$$

Interest Rate 5.00%

	single payment		Uniform payment series				Arithmetic Gradient	
n	F/P	P/F	A/F	A/P	F/A	P/A	A/G	P/G
4	1.2155	0.8227	0.2320	0.2820	4.3101	3.5460	1.4391	5.1028
5	1.2763	0.7835	0.1810	0.2310	5.5256	4.3295	1.9025	8.2369

Fig. 7.1. 5% interest rate table for periods 4 and 5.

Considering the information from the 5% interest rate table it will take between four and five periods. The number of periods, using the Excel function nper, is 4.57 periods.

Option 2:

$$PW = -\$100{,}000 + \$10{,}000\ \left({}^{P}/_{A}, 5\%, n\right)$$

$$\left({}^{P}/_{A}, 5\%, n\right) = \frac{\$100{,}000}{\$10{,}000} = 10$$

Interest Rate 5.00%

	single payment		Uniform payment series				Arithmetic Gradient	
n	F/P	P/F	A/F	A/P	F/A	P/A	A/G	P/G
14	1.9799	0.5051	0.0510	0.1010	19.5986	9.8986	5.7133	56.5538
15	2.0789	0.4810	0.0463	0.0963	21.5786	10.3797	6.0973	63.2880

Fig. 7.2. 5% interest rate table periods 14 and 15.

The number of periods for option 2 is between 14 and 15; the value from Excel is 14.20 periods. Therefore, option 1 should be selected since the initial investment is recovered more quickly.

In Example 7.3 we see that the number of periods for both alternatives is higher in comparison to the values obtained in Example 7.1 because of interest. Although this may seem obvious, this is one of the reasons paying credit card debt is so difficult, as interest rates are often above 20%.

Exercise

Calculate the payback period for the following alternatives considering a yearly interest rate of 7%.

	Cost	Yearly Savings
Alternative 1	$20,000.00	$3,500
Alternative 2	$35,000.00	$8,000

Answer Discount Payback = Alt 1 (7.55 years) I Alt 2 (5.4 years)

Benefit-Cost Ratio

A different approach to evaluate the financial viability of a project is to use a benefit-cost ratio analysis. This method assumes that a project is financially viable if it reaches the breakeven point where the benefits of a project are at least equal to its cost. In other words, an alternative is financially viable if the B/C ratio is greater or equal to 1. This requires a clear understanding of the following concepts:

- Benefit: Positive money inflow, such as profits and savings. For example, when buying an electric car, a benefit is the savings from buying electricity instead of buying gas.
- Cost: Money outflow of a project. Such as, the cost of the electric car. Private sector projects often consider only the initial cost; while the public sector often considers all costs, not just the initial cost.
- Disbenefit: Used for government or infrastructure projects. This amount does not represent a cost, but a reduction of benefits. For example, consider a diner that, due to building renovations, has reduced the number of seats available to customers, which is a disbenefit due to lost sales derived from the reduced capacity.

The equation for benefit cost ratio is described as follows:

Equation 7.2 Benefit-cost ratio

$$B/C = \frac{All\ benefits - disbenefit}{Cost} \geq 1$$

Example 7.4 shows a B/C calculation for Example 6.8.

Example 7.4

Determine the benefit cost for the following alternatives. Consider an interest rate of 5% per year and a life of 10 years for both alternatives.

Table 7.3 Data from Example 6.8

Option	Option 2	Option 1
Investment	\$100,000.00	\$10,000.00
Yearly return	\$15,000.00	\$2,500.00

For option 1 it shows the following:

$$B/C = \frac{\$2{,}500 * (P/A, 5\%, 10)}{\$10{,}000} = \frac{\$2{,}500 * (7.7217)}{\$10{,}000} = \frac{\$19{,}304}{\$10{,}000} = 1.9304$$

For option 2 it shows the following:

$$B/C = \frac{\$15{,}000 * (P/A, 5\%, 10)}{\$100{,}000} = \frac{\$15{,}000 * (7.7217)}{\$100{,}000} = \frac{\$115{,}825.5}{\$100{,}000} = 1.1582$$

Since both alternatives have a B/C ratio greater than 1, both options are financially feasible, and we perform a B/C of the increment.

$$Cost = \$100{,}000 - \$10{,}000 = \$90{,}000$$

$$Yearly\ return = \$15{,}000 - \$2{,}500 = \$12{,}500$$

$$B/C = \frac{\$12{,}500 * (P/A, 5\%, 10)}{\$90{,}000} = \frac{\$12{,}500 * (7.7217)}{\$90{,}000} = \frac{\$96{,}521.25}{\$90{,}000} = 1.0724$$

Because the B/C for the increment is greater than 1, the highest cost option 2 should be selected.

Example 7.5 describes a benefit-cost example for a public infrastructure project. In this example all benefits and costs are considered. It is important to note the disbenefits are subtracted from the benefits.

Example 7.5

A large airport has experienced an increase in demand in the past years. Forecasts indicate that the demand will continue to increase in the upcoming years. State officials see this as a fantastic opportunity to increase the state's competitiveness, so they want to use B/C ratio to evaluate the following plan.

The airport renovation would take five years and have an initial cost of $1 billion in year 0 and a cost of $800 million for the following years. This cost includes the addition of a new runway, updated taxiways and 40 additional gates so that more planes can be parked. Furthermore, this would also allow the construction of local roads and a new subway terminal.

IMG 7.3

This renovation would increase airport capacity to fifteen million passengers per year, from the initial capacity of 10 million. The airport fees per passenger average $50. During the renovation, the airport capacity would be reduced by 3 million passengers. Assume an interest rate of 10% per year and project duration of 25 years (5 years for the renovation and 20 years for cost recovery).

Cost in year 0: $1 billion

Yearly cost years 1 to 5: $800 million

Yearly benefit = number of passengers * yearly passenger fees

Yearly disbenefit = passenger reduction * yearly passenger fees

Periods (n) = 20 years

Interest rate = 10%

The income is calculated by adding the income from the renovation years and the income after the renovation and by subtracting the disbenefits from the renovation years:

$$Benefit = benefit - disbenefits$$

Yearly benefit years (1-5) = 10 M passengers * passenger fees = 10 M *$50 =$500 M

Yearly disbenefit years (1-5) = 3 M passengers * passenger fees = 3 M *$50 =$150 M

Net Yearly benefit years (1-5) = 7 M passengers * passenger fees = 7 M *$50 =$350 M

Yearly benefit after renovation = 15 M passenger * passenger fees= 15 M *$50 = $750 M

$$Benefit\ Equation$$

$$Benefit = \$350\,M\,(P/A, 10\%, 5) + [\$750\,M\,(P/A, 10\%, 15) * (P/F, 10\%, 5)]$$

$$Benefit = \$350\,M\,(3.791) + [\$750\,M\,(7.6061) * (0.6209)]$$

$$Benefit = \$1,326.78\,M + \$3,542.08 = \$4,868.86$$

Example 7.5 (Continued)

The cost is calculated by adding the cost in year 0 and the renovation cost from years 1 to 5:

$$Cost = \$1\,B + \$800\,M\,({}^{P}/{}_{A}, 10\%, 5)$$

$$Cost = \$1B + \$800\,M\,(3.7908) = \$1B + \$3{,}032.63M = \$4{,}032.63M$$

Therefore, the B/C is calculated using Equation 7.2:

$$^{B}/_{C} = \frac{\$4{,}868.86\ M}{\$4{,}032.63\ M} = 1.2073$$

Since the B/C is greater than 1 the renovation should be performed.

From examples 7.4 to 7.5 we learned the following:

1. The B/C ratio is a useful technique to determine when an alternative reaches a breakeven point.
2. When two or more alternatives have a B/C ratio higher than 1, an incremental analysis should be performed to determine the best alternative.
3. B/C ratio for public and infrastructure projects include all the benefits, disbenefits and costs.
4. Example 7.5 describes a B/C ratio greater than 1. However, some public infrastructure could be developed even when the ratio is lower than 1. This occurs when the public benefit surpasses the financial benefit. For instance, a park may not generate an immediate financial benefit but the public benefit in the form of a leisure space or to attract tourism may outweighing the financial cost.

Exercise

Calculate the B/C ratio for the following alternatives assuming a yearly interest rate of 7% and a life of 15 years for both alternatives.

	Cost	Yearly Savings	Disbenefit
Alternative 1	$20,000.00	$3,500	$10,000
Alternative 2	$35,000.00	$8,000	$15,000

Answer :
B/C Alt 1= 1.09
B/C Alt 2= 1.65

Future Worth Calculation

As previously described, PW is a useful way to select an investment alternative by learning which alternative has a greater value in today's money. However, FW calculations are especially useful for savings and retirement plans. FW calculations are performed by calculating the value of the different cash flows to a given future. Consider Example 7.6

Example 7.6

Marco likes to lease a car to maintain a recent model and to avoid the problem of buying and selling a vehicle. Therefore, he plans lease and to keep renewing his lease every three years for the following thirty-six years.

After adding signing fees and monthly payments he calculated his monthly payment as $850, Determine the FW of Marco's lease behavior. Use a yearly interest rate of 9%.

$i = 9\%\ yearly$

$i = 0.75\%\ monthly$

$n = 36\ years$

$n = 432\ months$

$Annuity = \$850$

Using Equation 3.3

$$F = \frac{A[(1+i)^n - 1]}{i}$$

$$F = \frac{850\ [(1+0.0075)^{432} - 1]}{0.0075}$$

$$F = \$850 * 3{,}230.25$$

$$F = \$2{,}745{,}713.97$$

When multiplying $850 for 432 months we obtained a value of $367,200, which is a value calculated with no interest, which is quite different to the value of $2,745,713.97. This shows the effect of long-term compounding interest and the relevance of early investing.

Exercise

A family has two options for summer vacation. One has a cost of $15,000 to a Florida amusement park and another going to different national parks with a cost of $5,000. Determine the future value for each alternative and the future value of the difference. Assume consider an interest rate of 9% per year and a time horizon of 15 years.

Answer :
15,000= $237,205.19
5,000= $79,068.40
Difference (10,000) =$158,136.79

Sensitivity Analysis

When mechanical engineers make a new machine or when civil engineers start a new project such as a bridge, they use a term called "safety factor," which is used to describe how much stronger a system is than it needs to be for an intended load. For example, a bridge load calculation can be set to support 1,000 tons and given a safety factor of 30% it will be designed and built to support a load of 1,300 tons.

Engineering economy projects are no different. We may modify our values to answer the question of What if? as described in Example 7.7.

Example 7.7

A civil engineering company is planning a new toll bridge. The bridge will likely transport 50,000 motorists each day. Determine what would occur if the number of motorists increases and decreases by 20%. The cost of the project is $250M determine the cost of the toll assuming that they want to recover their investment in 10 years. Use an interest rate of 7% per year.

Present value = $250 million

Motorists per year = 50,000 * 365 = 18,250,000

For simplicity, using yearly capitalization.

$$PW = -\$250{,}000{,}000 + toll * 18{,}250{,}000(P/A, 7\%, 10)$$

$$Toll = \frac{\$250{,}000{,}000}{18{,}250{,}000 * 7.024} = \$1.95$$

Motorists	18,250,000	+ 20% motorists 21,900,000	− 20% motorists 14,600,000
Toll	$1.95	$1.63	$2.44

Therefore, the price of the toll may increase or decrease based on the number of motorists. Similarly, we may be able to add other values to the different parameters, such as maintenance cost, recovery period, or interest rate.

It should be noted that one of the reasons of the mortgage crisis of 2008 was that many borrowers considered an optimistic approach when acquiring a mortgage and did not perform an adequate sensitivity analysis about the what if, the mortgage payment increase? What if the loan amount is higher than the value of the house after two years? This, and many other questions, should be asked to be prepared for an eventuality.

Breakeven Analysis

A breakeven analysis equates two financial equations to a point in which one parameter equates both. Frequently this parameter is related to cost or duration. For example, in the case of cost companies would like to find the point where total cost and total revenue are equal. In addition to this perspective, in engineering economy we also equate interest rates as show in chapter 6 and we now consider the case of duration to find the number of periods where two financial alternatives are equal, such as the comparison between renting and buying as shown in example 7.8.

Example 7.8

Jules rents a two-bedroom apartment for a monthly payment of $2,500. He could also buy an apartment for a $60,000 down payment and a $1,500 monthly payment, which includes property taxes, maintenance costs, and mortgage payment. Use a PW analysis to determine the number of periods that would make both alternatives equal consider a 5% yearly interest rate.

For simplicity we consider yearly compounding interest.

Yearly rent cost is $2,500 * 12 = $30,000

Yearly mortgage cost is $1,500*12 = $18,000

PW rent

$$PW = \$30{,}000 * ({}^{P}/_{A}, 5\%, n)$$

PW buying

$$PW = \$60{,}000 + \$18{,}000 * ({}^{P}/_{A}, 5\%, n)$$

Equating both equations

$$({}^{P}/_{A}, 5\%, n) = \frac{\$60{,}000}{\$12{,}000} = 5$$

Interest Rate 5.00%

	single payment		Uniform payment series				Arithmetic Gradient	
n	F/P	P/F	A/F	A/P	F/A	P/A	A/G	P/G
5	1.2763	0.7835	0.1810	0.2310	5.5256	4.3295	1.9025	8.2369
6	1.3401	0.7462	0.1470	0.1970	6.8019	5.0757	2.3579	11.9680

Fig. 7.3. 5% interest rate table periods 5 and 6.

Therefore, the value number of years is between 5 and 6 years. We could interpolate to solve the number of periods, using the nper function in Excel with parameters NPER (5%,-12000,60000), which generates a solution of 5.89 periods, as seen in Fig. 7.4.

Example 7.8 (continued)

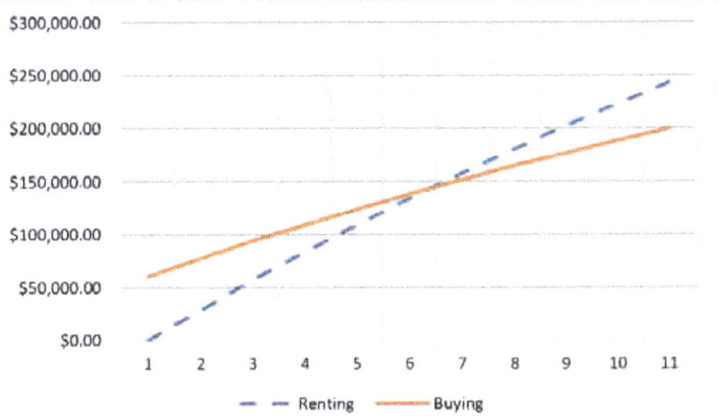

Fig. 7.4. Breakeven analysis for renting and buying.

Uncertainty

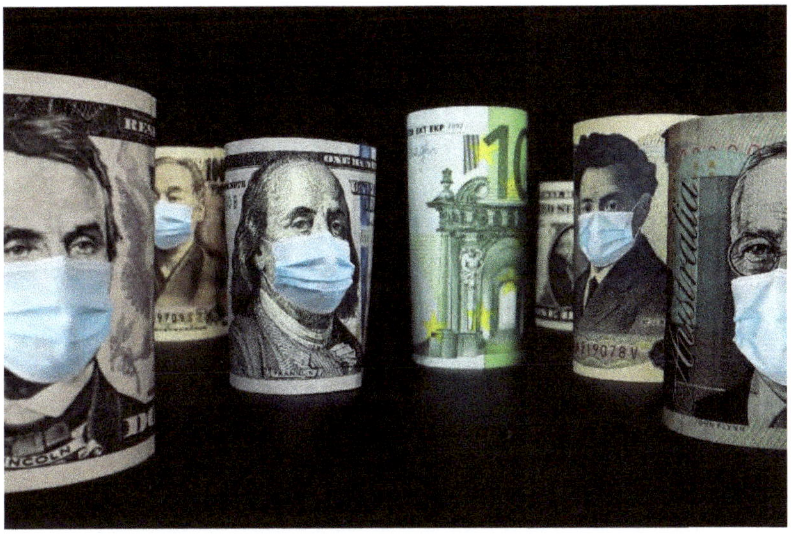

IMG 7.4

This generation has been marked by uncertainty. In 2008 the housing market collapsed; in 2020 the world experienced the COVID-19 pandemic and the Russian invasion to Ukraine. Smaller events related to the force of nature, such as draughts, hurricanes and earthquakes can have negative effects. Similarly, changes in economic policy can bring negative outcomes such as the case of Brexit for the UK, or mostly positive as in the case of NAFTA now called USMCA. According to the office of the US Trade Representative [1] since the creation of NAFTA U.S. exports to Mexico are up 517% from 1993 (pre-NAFTA) and are also up 191% to Canada from 1993 (pre-NAFTA)".

[1] Mexico. "Office of the United States Trade Representative." https://ustr.gov/countries-regions/americas/mexico 12/10/2022

Therefore, we know that the only thing that is constant in this world is uncertainty, and we can either choose to ignore it or to embrace it as part of the equation. Unfortunately, ignoring uncertainty may not be a wise choice as we will be unprepared when it arises. Consequently, this section describes how to consider uncertainty in a financial project.

There are four main approaches to deal with uncertainty:

- Estimation Methods
- Expected Values
- Decision Trees
- Risk analysis

Estimation for Uncertainty

When starting a new project, the amount of information available is limited. For example, in the introductory reading Diego has historical information about his sales volume for his current customers in the entertainment business but lacks sales information for his expansion. Furthermore, different customers may have different preferences, so a high-selling product may not be as successful in the other market. In cases like this, with little information it is necessary to generate an estimate.

One way to generate parameters is to create a range of possible values, that represent three different approaches, an optimistic value, a most likely value, and a pessimistic value, and can be used to calculate the range of rates of return for an investment alternative that can in turn be compared with the MARR, as shown in Example 7.9.

Example 7.9

As described, Diego will expand his business. The following table describes the investment values, income values, and duration values. If Diego has a MARR of 14%, determine the project IRR for the different perspectives and if the project should be performed.

	Optimistic	Most likely	Pessimistic
Investment	\$50,000	\$65,000	\$80,000
Income	\$13,118	\$17,175	\$22,745
Duration	7	6	5

Equation for the optimistic estimate

$$PW = -\$50{,}000 + \$13{,}118\left(P/A, i, 7\right)$$

$$\left(P/A, i, 7\right) = \frac{\$50{,}000}{\$13{,}118} = 3.8115$$

$$i = 18\%$$

Example 7.9 (Continued)

Equation for the most likely estimate:

$$PW = -\$65{,}000 + \$17{,}175\,(P/A, i, 6)$$

$$(P/A, i, 6) = \frac{\$65{,}000}{\$17{,}175} = 3.7844$$

$$i = 15\%$$

Equation for the pessimistic:

$$PW = -\$80{,}000 + \$22{,}745\,(P/A, i, 5)$$

$$(P/A, i, 7) = \frac{\$80{,}000}{\$22{,}745} = 3.5172$$

$$i = 13\%$$

Since the MARR is 14% and the pessimistic approach is lower than 14%, Diego may decide not to move forward with this investment.

A different approach is to use the range of values previously described to calculate an estimate for each parameter following a beta approximation and then determine the IRR. The beta approximation is a weighted average described in Equation 7.3. Example 7.10 presents this approach.

Equation 7.3 Beta approximation

$$Estimate = \frac{(Pessimistic\ value * 1) + (Most\ Likely\ value * 4) + (Optimistic\ value * 1)}{6}$$

Example 7.10

Determine the IRR for the data in Example 7.9; use Equation 7.3 to determine the estimates for investment, cost, and duration. Consider a MARR of 14%.

	Optimistic	Most likely	Pessimistic
Investment	$50,000	$65,000	$80,000
Income	$22,745	$17,175	$13,118
Duration	7	6	5

Investment estimate:

$$Estimate = \frac{(\$50{,}000 * 1) + (\$65{,}000 * 4) + (\$80{,}000 * 1)}{6} = \$65{,}000$$

Example 7.10 (Continued)

Income estimate:

$$Income\ Estimate = \frac{(\$22{,}745 * 1) + (\$17{,}175 * 4) + (\$13{,}118 * 1)}{6} = \$17{,}427.48$$

Duration estimate:

$$n\ Estimate = \frac{(7 * 1) + (6 * 4) + (5 * 1)}{6} = 6$$

Therefore, the PW calculation is estimated as follows:

$$PW = -\$65{,}000 + [\$17{,}427.48 * (P/A, i, 5.17)]$$

$$(P/A, i, 7) = \frac{\$65{,}000}{\$17{,}427.48} =$$

$$i = 15.54\%$$

Since the MARR is 14%, Diego should perform the project as the estimated value is higher than the MARR.

Estimation for Uncertainty with probability

IMG 7.5

Every business face uncertainty. The initial reading explains that Diego had been working for the past 10 years in the food truck business. This means that Diego has historical information he can use to generate forecasts or to define the likelihood of an event. However, this does not mean that every event should be considered for a probability analysis and only relevant variables should be considered for a probability analysis. For example, Diego could calculate the probability of certain events such as reaching a certain sales volume.

Before we continue with the application of probabilities in an engineering economy there are some probability rules that need to be considered:

1. Probability cannot be negative or greater than 1.
2. The sum of all probabilities is equal to 1.

In addition, we should note that the nature of the data can be deterministic or stochastic. In simple terms we can describe deterministic data as "known" information that can considered as constant. For example, when we performed a PW in Chapter 4 for different lives Example 4.5, we determined that the cost of renewal was the same as the original cost. Stochastic data is "unknown" data that can follow a probability distribution. For example, in the case of Diego, for the number of beverages sold, he may know the highest, lowest, and average sales for the past month, but he does not know how many beverages will be sold tomorrow.

Similarly, to deterministic data, deterministic probabilities assume known probabilities. For example, Diego may have a 20% probability of selling \$10,000. In engineering economy, it is common to consider discrete probabilities, with a range of two to five different probabilities and parameters, such as cost, annuity, or future value.

Once the probabilities are known for the different parameters, an expected value can be obtained by multiplying each parameter by its corresponding probability, as described in Equation 7.4.

Equation 7.4 Expected value

$$Expected\ value = (Parameter\ 1 * Probability\ 1) + \cdots + (Parameter\ n * Probability\ n)$$

Example 7.11

Determine the expected value for the investment data from Example 7.9 considering the probabilities described.

Investment	Probability
\$50,000	30%
\$65,000	45%
\$80,000	25%

$$Expected\ value = (Parameter\ 1 * Probability\ 1) + \cdots + (Parameter\ n * Probability\ n)$$

$$Expected\ value = (\$50{,}000 * 30\%) + (\$65{,}000 * 45\%) + (\$80{,}000 * 25\%)$$

$$Expected\ value = \$64{,}250$$

Joint Probabilities

Consider a situation in which one or more parameters have a probability. In such a situation, calculate the joint probabilities from the different parameters.

To exemplify this situation, consider the flip of a fair coin; in the first flip there is a 50-50 chance for each side. Similarly, the following flips will also have a 50-50 chance for each side. In order to calculate the chance of getting two consecutive flips we would multiply the probability of getting heads in each flip, as seen in Fig. 7.5.

Therefore, we would get a probability of 25%.

$$p = \frac{1}{2} * \frac{1}{2} = \frac{1}{4} = 25\%$$

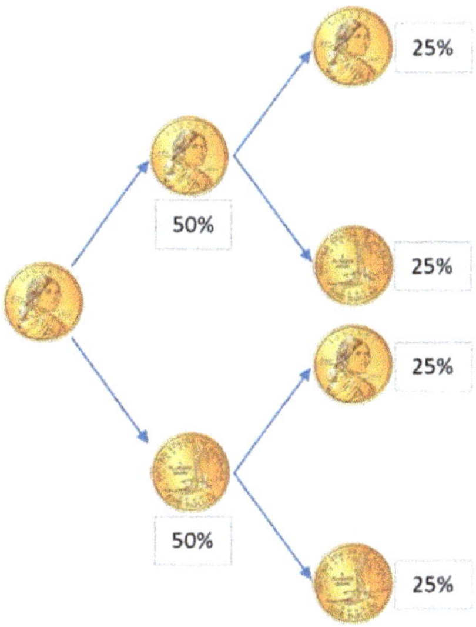

Fig. 7.5. Joint probabilities for a coin toss.

Example 7.12

Determine the expected value and the PW value from Example 7.10 assuming the cost and probabilities described, a yearly interest rate of 10%. The project can have a life of 5 years with a probability of 20%, a life of 6 years with a probability of 65%, or a life of 7 years with a probability of 15%.

Investment	Probability	Income	Probability	Life	Joint Probability	PW	PW * Joint Probability
$50,000	30%	$7,143	20%	5	6%	-$3,969.02	-$238.14
$65,000	45%	$7,143	20%	5	9%	-$13,282.84	-$1,195.46
$80,000	25%	$7,143	20%	5	5%	-$22,596.66	-$1,129.83
$50,000	30%	$10,833	65%	6	20%	$18,958.29	$3,696.87
$65,000	45%	$10,833	65%	6	29%	$10,491.19	$3,068.67
$80,000	25%	$10,833	65%	6	16%	$2,024.08	$328.91
$50,000	30%	$16,000	15%	7	5%	$52,236.80	$2,350.66
$65,000	45%	$16,000	15%	7	7%	$44,539.42	$3,006.41
$80,000	25%	$16,000	15%	7	4%	$36,842.05	$1,381.58
				SUM	100%	$125,243.31	$11,269.67

The joint probability is for the $50,000 investment, and the income of $7,143 is obtained by multiplying.

$$30\% * 20\% = 6\%$$

The remaining joint probabilities are calculated following a similar calculation.

The PW is calculated as

$$\text{PW} = -\$50{,}000 + \$7{,}143\left({}^{P}/_{A}, 10\%, 5\right) = -\$3{,}969.02$$

As described in the table the PW is multiplied by the joint probability to determine the expected value (EV), and the total EV for the project is calculated by adding the EVs from all investments and incomes.

Economic Probability Trees

Example 7.11 described an investment situation in which Diego was considering an investment project with different values and probabilities for cost and income. Unfortunately, investment projects are frequently more complex and often require a more detailed analysis in the form of a series of independent events that may include sequences and probabilities.

The economic probability tree is drawn using a sequence of nodes and branches in which each node (represented by a circle) represents an event and is associated with the probability of that event, and the branches (represented by an arrow) connects the different nodes. Branches are allowed to display alternative information and probability values. The probability for all branches must add to 100%.

Having an economic probability tree allows for the analysis of multiple alternatives, probabilities, and courses of action. The objective is then to generate an expected value for a multiple alternative problem.

To solve an economic probability tree, we need to completely the following steps:

1) Generate a problem statement.
2) Generate a tree representation.
3) Calculate the expected values for each branch using PW or EUAW.
4) Identify and select the branches with higher PW or EUAW and reject branches with lower PW or EUAW.
5) Calculate the EV for each branch.
6) Generate a final assessment by determining the EV for the project, considering each branch and its probabilities.

Example 7.13

Generate the economic probability tree and determine the EV for the following project. Consider a yearly MARR of 10% and a project life of 15 years. A company wants to analyze if they should invest in a new resort with a cost of $300 million.

The company is considering a six-month period for their calculations. After building the hotel they will have a three-semester period to analyze the hotel results and then decide their next step. They will consider discrete values for their analysis.

If their income reaches the optimistic value (option A shown in blue), they can expand their operations with an additional investment of $100 million, and in which they could either increase their income to reach $35 million per period with a probability of 65% or increase their income to reach $45 million per period. At the end of the 15 years of operation they may be able to sell the resort for $450 million for both probabilities.

If their income reaches the most likely income (option B shown in brown) it will generate $20 million per period with an estimated probability of 50%, and they may continue to operate without any further investment and sell after 15 years for $350M.

If their income reaches the pessimistic value (option C shown in green) it will generate $10 million per period. In such case the company has two options cease operations and sell for $200 million after the six-month period. They can also invest an additional $50 million to attract a different market and to generate an income of $30 million per period with a salvage value of $250 million at the end of the 15 years this option has a probability of 40%, or they could continue to operate with an income of $25 million per period for a period of 15 years with a salvage value of $200 million.

Fig. 7.6 shows the economic probability tree. We now proceed to calculate the PW for each branch, starting from the top, the optimistic approach, and the last two nodes, meaning after the $100 investment having an annuity and a salvage value.

Example 7.13

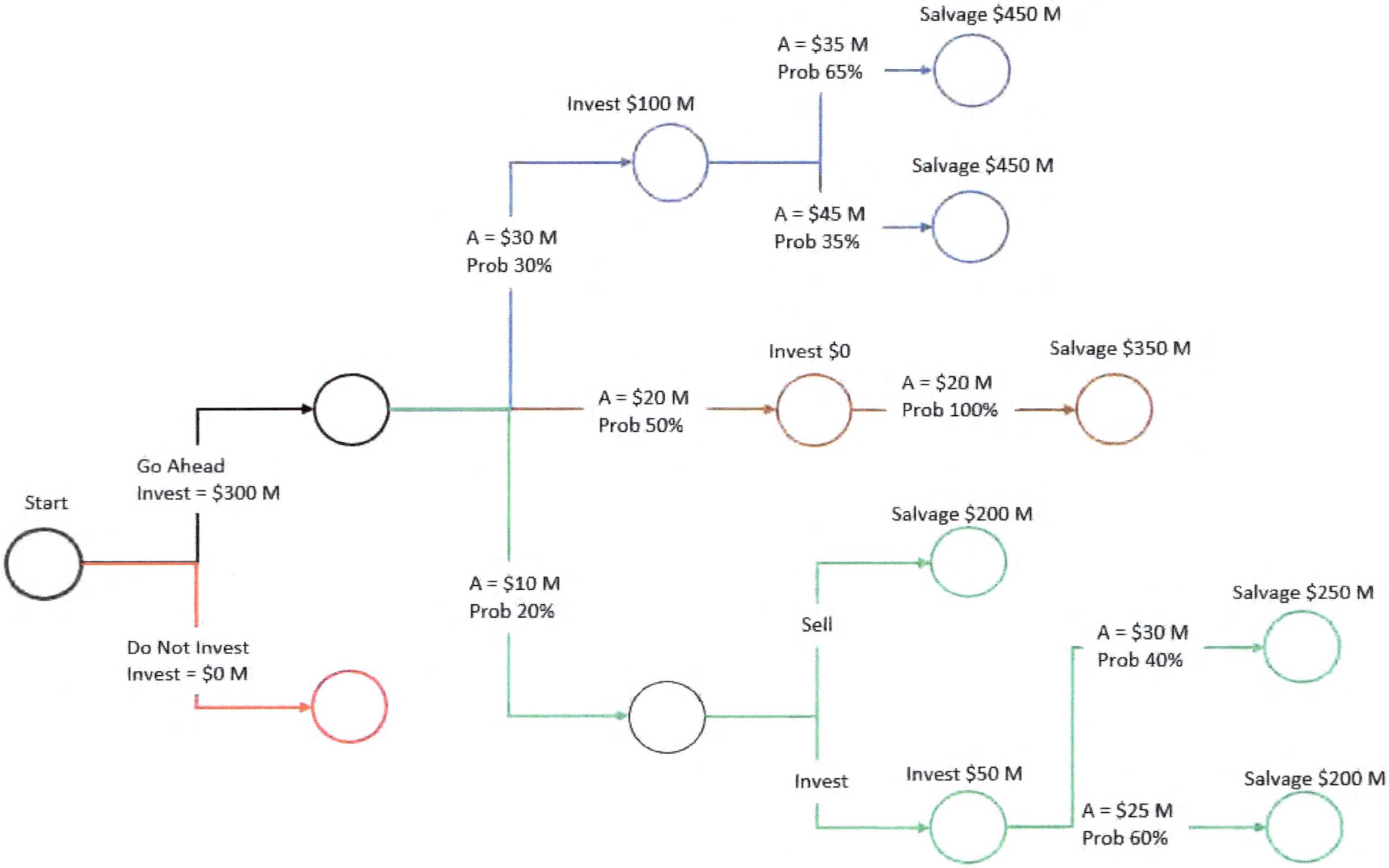

Fig. 7.6. Economic probability tree for Example 7.13.

Example 7.13 (Continued)

n=13.5 years or 27 semesters, i= 10% per year = 5% semester

Option A the optimistic approach:

Investment =$100 M, Annuity=$35 M, Salvage= $450 M

$$PW = -\$100\ M + \$35\ M(P/A, 5\%, 27) + \$450\ M(P/F, 5\%, 27) = \$533{,}037{,}920.25$$

Investment =$100 M, Annuity=$45 M, Salvage= $450 M

$$PW = -\$100\ M + \$45\ M(P/A, 5\%, 27) + \$450\ M(P/F, 5\%, 27) = \$679{,}468{,}256.45$$

Option B the most likely approach:

Investment =$0, Annuity=$20 M, Salvage= $350 M

$$PW = \$20\ M(P/A, 5\%, 27) + \$350\ M(P/F, 5\%, 27) = \$386{,}607{,}584.05$$

Option C the pessimistic approach:

Sell after 18 months.

$$PW = \$200\ M$$

Investment =$100 M, Annuity=$45 M, Salvage= $450 M

$$PW = -\$50\ M + \$30\ M(P/A, 5\%, 27) + \$250\ M(P/F, 5\%, 27) = \$556{,}253{,}088.35$$

$$PW = -\$50\ M + \$25\ M(P/A, 5\%, 27) + \$200\ M(P/F, 5\%, 27) = \$281{,}787{,}302.58$$

	Optimistic (A)	Optimistic (B)	Most Likely	Pessimistic (A)	Pessimistic (B)
Investment (PV)	-$100,000,000	-$100,000,000	$0	$50,000,000	$50,000,000
Annuity	$35,000,000	$45,000,000	$20,000,000	$30,000,000	$25,000,000
Salvage (FV)	$450,000,000	$450,000,000	$350,000,000	$250,000,000	$200,000,000
semesters (n)	27	27	27	27	27
interest rate (n)	5%	5%	5%	5%	5%
PW	$533,037,920.25	$679,468,256.45	$386,607,584.05	$556,253,088.35	$469,645,504.30
Probability	65%	35%	100%	40%	60%
Expected Value	$346,474,648.16	$237,813,889.76	$386,607,584.05	$222,501,235.34	$281,787,302.58

After adding optimistic and most likely values the values per alternative are:

Expected value	Optimistic	Most Likely	Pessimistic
	$584,288,537.92	$386,607,584.05	$504,288,537.92

Example 7.13 (Continued)

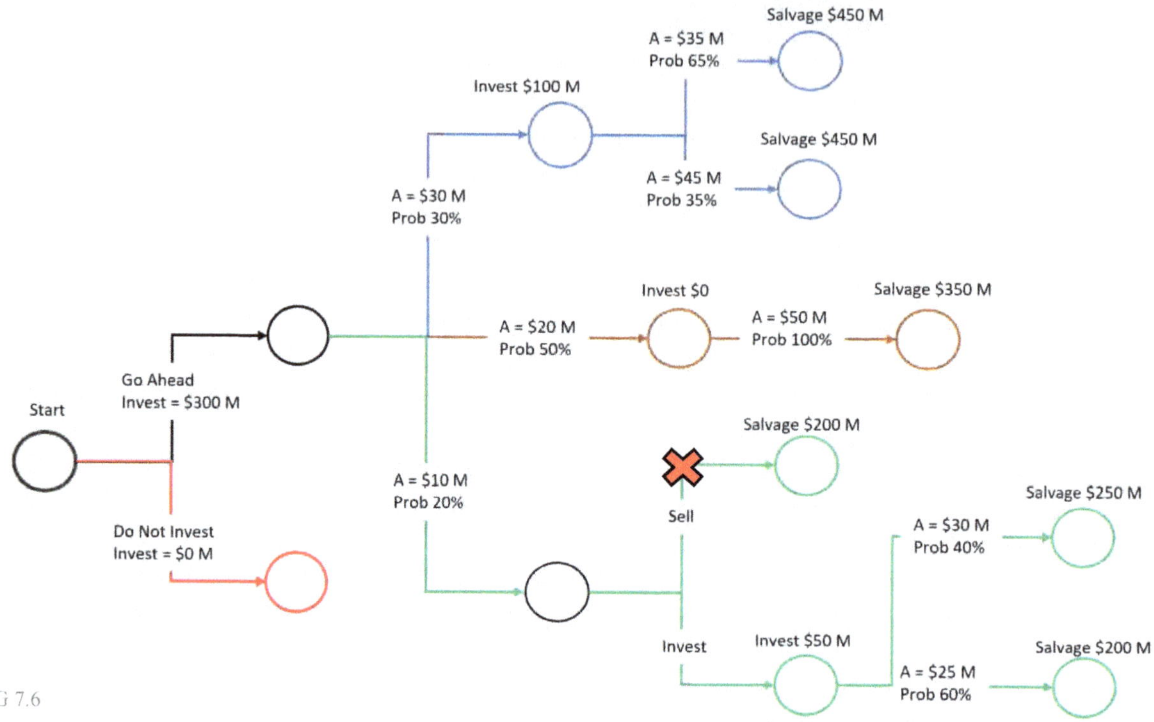

IMG 7.6

Therefore, from the previous results, the optimistic value from the two branches has a value of \$584,288,537.92; the most likely value does not offer any options and has a value of \$386,607,584.05 and the pessimistic value should not consider the sell option as it has a value of \$200,000,000 which is far less than \$504,288,537.92.

The expected PW from the three branches by multiplying the value of each branch by its probability.

$$PW = (584{,}288{,}537.92 * 30\%) + (\$386{,}607{,}584.05 * 50\%) + (504{,}288{,}537.92 * 20\%)$$

$$PW = \$469{,}448{,}060.98$$

Since this PW is the PW for the next 27 periods, we need to determine the expected value for the first three annuities to determine the PW in time 0:

$$EV = (\$30\text{ M} * 30\%) + (\$20\text{ M} * 50\%) + (\$10\text{ M} * 20\%) = \$21{,}000{,}000$$

Considering the initial investment, the annuity from the first three periods and the EV for the following 27 periods as shown below.

$$PW = -\$300\,M + \$21\,M(P/A, 5\%, 3) + \$469{,}448{,}060.98(P/F, 5\%, 3)$$

$$PW = -\$300\,M + \$57{,}188{,}208.62 + \$405{,}526{,}885.64 = \$162{,}715{,}094.25$$

Therefore, the go-ahead option should be selected and the investment will generate a positive PW of \$162 million.

Economic probability trees are also used for simpler problems as shown in example 7.14

Example 7.14

Alberto is renting a new apartment and has the option to get renter's insurance or to take the risk and of not insuring his apartment. Renter's insurance has a cost of $500 per year with a $50 deductible. Alberto lives in an area that is not very safe, so there is a 10% probability of suffering a small robbery and a 5% probability of suffering a big robbery, and the remaining percentage of 85% of not suffering a robbery. The small robbery has a cost of $1,000, and the large robbery has a cost of $4,000.

Draw the economic decision tree, determine the expected value, and decide if Alberto should get renter's insurance.

The economic decision tree is as follows:

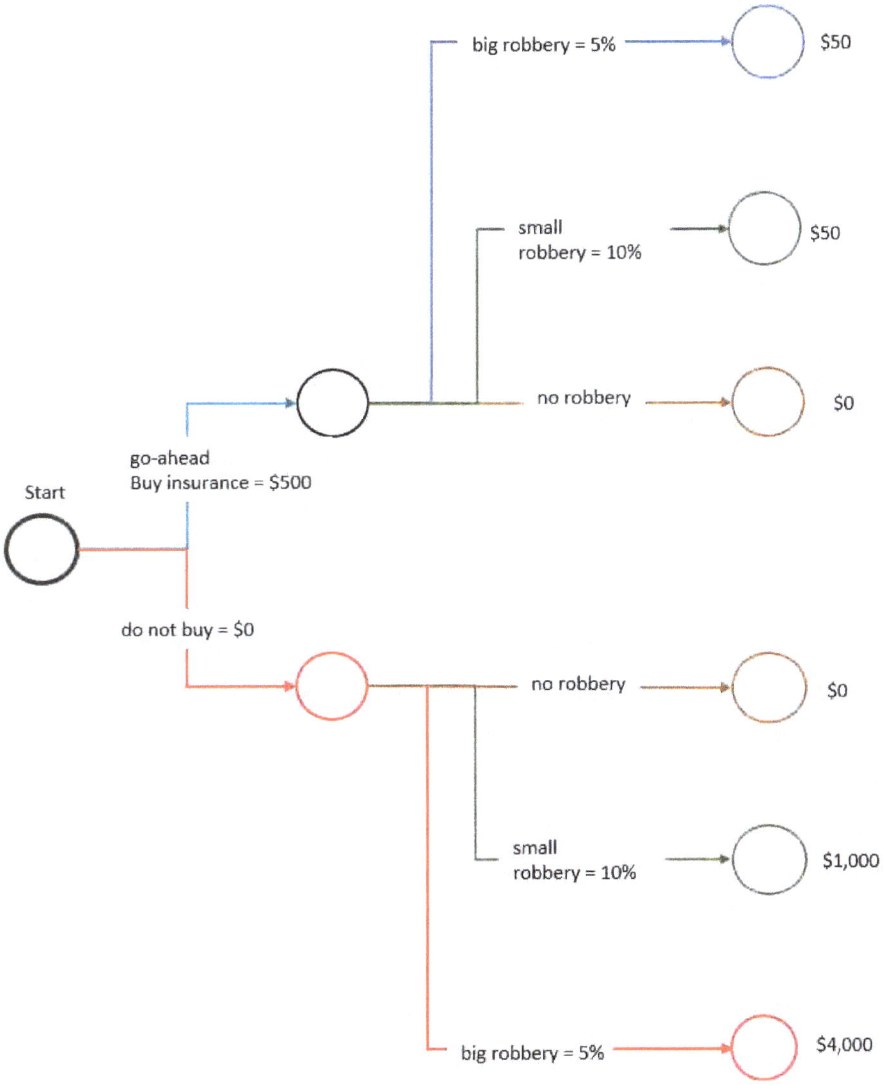

Fig 7.7. Economic probability tree for Example 7.14.

Example 7.14 (Continued)

The EV for buying insurance is.

$$EV = (\$50 * 5\%) + (\$50 * 10\%) = \$7.5$$

Therefore, the cost of buying insurance is \$500 + \$7.5 = \$507.5

The EV for not buying insurance is.

$$EV = (\$4,000 * 5\%) + (\$1,000 * 10\%) = \$300$$

Therefore, if Alberto were to buy insurance, he would spend \$207.50 more than not buying insurance. On the other hand, Alberto may want to limit his loss to \$507.5 instead of experiencing a potential loss of \$4,000.

IMG 7.7

Risk Management

Expected values are relevant because they allow us to incorporate probability information into an economic analysis. However, they should not be the only consideration. For example, buying a very affordable house in an area with a high flooding risk may not be a clever idea, even if the EV is positive. In other words, risk will allow us to incorporate in the equation the chance of getting an outcome different that the one described by the EV. We should note that risk has a negative connotation so we are considering the possibility of negative effects and not positive effects, which can also happen unexpectedly.

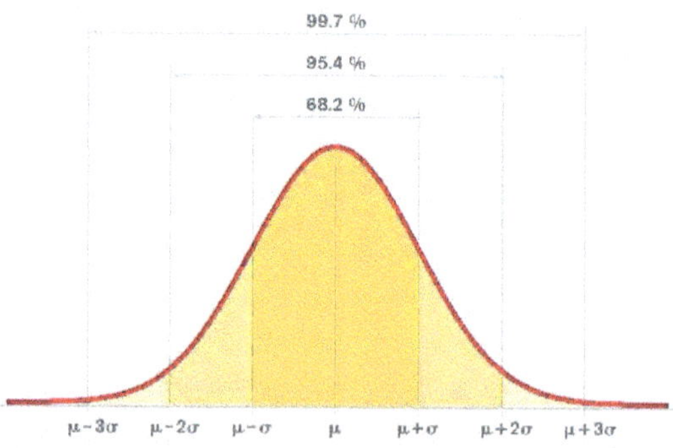

Fig. 7.8. Normal distribution with $\pm 3\sigma$ standard deviations.

One way to consider risk is to assume that risk follows a normal distribution. As shown in Fig. 7.8, $\pm 1\sigma$ standard deviation considers 68.2% of the probable outcomes, $\pm 2\sigma$ standard deviations consider 95.4% of the probable outcomes, and 99.7% of probable outcomes fall within $\pm 3\sigma$ standard deviations. A standard deviation represents the squared difference between a data point and the mean value of the data.

In quality control the goal is to provide consistency; for instance, when filling a water bottle, we want the content to fall within a parameter. If the standard deviation is high, it means that the water content has a high variability and is not consistent. Following this approach, we can calculate the standard deviation from the expected value EV as

Equation 7.4 Standard deviation for EV

$$Standard\ deviation\ (\sigma) = \sqrt{[EV(X-\bar{x})^2]}$$

The following equation is equivalent to Equation 7.4 and considers the probability of each outcome:

Equation 7.5 Square root of the difference between the average of the squares and the square of the average

$$Standard\ deviation\ (\sigma) = \sqrt{\{EV(X^2) - [EV(X)]^2\}}$$

$$Standard\ deviation\ (\sigma) = \sqrt{\{Outcome_1^2 * P(1) + Outcome_2^2 * P(2) + \ldots + Outcome_n^2 * P(n) - EV^2\}}$$

Example 7.13 shows the risk calculation for Example 7.12.

Example 7.15

Consider the data and results from Example 7.12.

The EV from buying insurance is.

$$EV = (\$50 * 5\%) + (\ \$50 * 10\%) = \$7.5$$

Therefore, cost of buying insurance is \$500 + \$7.5 = \$507.5.

The EV from not buying insurance is.

$$EV = (\$4{,}000 * 5\%) + (\ \$1{,}000 * 10\%) = \$300$$

First, considering the second equation,

$$(\sigma) = \sqrt{\{Outcome_1^2 * P(1) + Outcome_2^2 * P(2) + \ldots + Outcome_n^2 * P(n) - EV^2\}}$$

buying insurance $(\sigma) = (0.85 * 0^2) + (0.1 * \$50^2) + (0.05 * \$50^2) = \$375$

NOT buying insurance $(\sigma) = (0.85 * 0^2) + (0.1 * \$1000^2) + (0.05 * \$4000^2) = \$900{,}000$

With this information we can consider the first part of Equation 7.5.

Example 7.15 (Continued)

$$Standard\ deviation\ (\sigma) = \sqrt{\{EV(X^2) - [EV(X)]^2\}}$$

Buying insurance $(\sigma) = \sqrt{\$375 - \$7.5^2} = \sqrt{\$318.75} = \$17.85$

Not buying insurance $(\sigma) = \sqrt{\$900{,}000 - \$300^2} = \sqrt{\$810{,}000} = \$900$

We see that even when the EV from not buying insurance was $300, it is lower than the EV cost from acquiring insurance, $507.5.

On the other hand, the standard deviation from **not** buying insurance is more than 50 times larger, which is far higher and riskier. Could it be that someone doesn't buy the insurance and gets results? Yes, it possible however, it is riskier, and it depends on the level of risk that a person or company wants to take on.

What Did We Learn in This Chapter?

1. We learned new methods to assess the financial viability of a project.
2. We learned how to calculate the benefit-cost ratio for private and government projects.
3. We learned how to determine the number of periods required to recover an investment with and without interest rates.
4. We learned how to calculate FW calculations for investment projects.
5. We learned how to equate two investment alternatives in order to find a breakeven analysis.
6. We learned that uncertainty affects engineering economy projects and how to determine estimates following a beta approximation and considering probabilities.
7. We learned how to construct economic probability trees and how to estimate risk.

Review Questions

Directions: Refer to what you learned in this chapter in order to answer the following questions:

1. What is the difference between Benefit-Cost ratios for private projects and for government projects? Should they be considered differently?
2. What method does Kevin O'Leary use when assessing payback periods? What is the advantage of this method?
3. Consider a situation in which you have two alternatives: one with a PW value of $100,000 and another with a value of $105,000. When calculating FW for both options, is it possible that the one with the PW of $100,000 has a higher FW than the other? If so, why?
4. When should you use a beta approximation to calculate an estimate?
5. In a new project, should you consider using the beta approximation or the historic data to determine probabilities?
6. When solving an economic probability tree, do we start from left to right or vice versa? Why?
7. Why do we use the standard deviation for risk calculation?

Chapter 7 Practice Problems

1. Computer the future worth for the following cash flow in year 6. Assume a yearly interest rate of 12% per year.

Year	0	1	2	3	4	5	6
Flow	\$500	\$600	\$700	\$800	\$900	\$0	F

2. Computer the future worth for the following cash flow in year 4. Assume a yearly interest rate of 12% per year.

Year	0	1	2	3	4
Flow	\$10a	\$8a	\$6a	\$4a	\$2a

3. A local company invested for 25 years using monthly compounding at a 9% interest rate with the goal to reach \$100 million.
    a. Determine how much they should invest each monthly period.
    b. Assume the value of the annuity from part a as an initial annuity. Determine the amount of their savings after 25 years if they increase their savings by 2% each month.
4. Computer the future worth for the following cash flow in year 13. Assume a yearly interest rate of 12% per year.

Year	0	1	2	3	4	5	6	7	9	10	11	12	13
Flow	\$500	\$600	\$700	\$800	\$900	\$900	\$900	\$900	\$900	\$900	\$900	\$900	\$900

5. The following mutually exclusive options have a life of seven years. Assume a MARR of 12%. Determine the IRR for each project and which project should be selected.

	Alternative A	Alternative B	Alternative C
Cost	\$6,000.00	\$5,000.00	\$4,500.00
EUAB	\$1,356.66	\$993.45	\$1,081.62

6. The following mutually exclusive options have a life of seven years. Assume a MARR of 12%. Determine the IRR for each project and which project should be selected.

	Alternative A	Alternative B	Alternative C
Cost	$6,000.00	$5,000.00	$4,500.00
EUAB	$1,356.66	$993.45	$1,081.62
Salvage value	$500.00	$300.00	$1,000.00

7. A local government is considering building a new park. There are four different locations. The cost of the land and their operation and maintenance cost are provided.

	Alternative A	Alternative B	Alternative C	Alternative D
Cost	$ 60 M	$50 M	$70 M	$40 M
Yearly operation and maintenance	$1.5 M	$2.2 M	$8.2 M	$3.2 M

Assume a benefit-cost ration to determine which alternative should be selected. Assume that all parks will provide a health service that is estimated to be $6.5 million. The government uses a 7% interest rate per year.

8. Determine which of the following alternatives should be selected using the payback period.

	Year 0	Year 1	Year 2	Year 3	Year 4	Year 5	Year 6	Year 7
Alternative A	-$50,000	$5,000	$5,000	$10,000	$10,000	$20,000	$20,000	$70,000
Alternative B	-$65,000	$2,500	$7,000	$5,000	$15,000	$10,000	$12,000	$70,000

9. Assume the following alternatives with a life or eight years with no salvage value and a MARR of 9%. Determine the best alternative using the following methods:

   a. The payback period and the discounted payback period

   b. The future worth

   c. Benefit-cost ratio

	Alternative A	Alternative B	Alternative C	Alternative D
Cost	$ 60 M	$50 M	$70 M	$25 M
Savings per period	$14 M	$13 M	$18.2 M	$7 M

10. A company uses a MARR of 10% per year and wants to determine the value of X periods that will make the investment alternatives equal:

	Alternative A	Alternative B
Cost	$ 60 M	$50 M
Savings	$10 M	$8 M
Periods	12	X

11. Smock is a local computer company that provides a virtual server service for small companies. They just acquired a new server, and the installation and setup are expected to take 45 days. The installation time depends on the availability of different outside contractors, who are currently finishing a previous contract. Based on previous records, they are most likely to be available in 30 days. However, they could be available as early as 20 days or as late as 55 days. Determine the estimated waiting time.

12. Over the last decade, robotic-drone research and development company researched new drone technology for home delivery. Their sales department thinks that they could sell as many as $90 million per year with a probability of 20%. However, their sales most likely value is projected to reach $76 million with a probability of 45% and their lowest value sales could go for as little as $50 million.

    a. Determine the expected value of their sales.

    b. The sales department estimates that they could maintain this sales rhythm for four years with a probability of 67% or that this sales rhythm can be maintained for eight years. Determine the expected value for the PW if the company Assumes a yearly interest rate of 10%.

13. Paul wants to invest in a new business. The initial investment is $10,000, and they are looking at three different marketing campaigns:

    a. The first campaign could have three different outcomes with the following values: a high value of $15,000 with a probability of 35%, a most likely outcome of $10,000 with a probability of 40%, and a least likely outcome with a value of $8,000.

    b. The second campaign could have two different outcomes: a high value of $22,000 with a probability of 60% and a low value of $12,000.

    c. The third campaign could have two different outcomes: a low value of -$5,000 with a probability of 20% and a high value that must be calculated for the following description.

        i. The high value comes from two possibilities, each with 50% probability.

            1. The first possibility comes from two alternatives; the first alternative has a value of $12,000. With a probability of 20% and the second alternative has a value of $25,000

            2. The second possibility comes from two alternatives; the first alternative has a value of $15,000 with a probability of 35% and the second alternative has a value of $25,000.

14. A company hires you to perform a risk analysis for flood insurance for a building with a value of $2 million. Since flooding can reach various levels, it could generate different levels of loss:

Flooding loss	Probability
No flooding loss	0.8
$20,000	0.1
$30,000	0.05
$40,000	0.03
$50,000	0.02

a) Determine the expected flood loss in any year.

b) If the expected flood loss is less than the cost of $4,000, determine why the company still should buy flood insurance?

Chapter 8 Depreciation and Taxes

Diego's Taxes

IMG 8.1

As described in the previous chapter Diego, Paula's brother, expanded his food truck business: he acquired a new food truck and some additional equipment for $180,000.

His investment included a portable solar plant with a value of $10,000 to reduce the need for traveling with additional fuel to generate electricity.

Diego was unsured about how to calculate his income tax and asked his friend George for help. Last year Diego's company achieved a new sales record of $1,000,000 with only $55,000 monthly expenses. One of the multiple questions that Diego had was related to depreciation, when he acquired the solar plant, the salesperson said that the plant qualified for a 40% bonus depreciation.

Does this mean that he can depreciate the cost 40% in the first year and the remaining 60% in the following years? If so, would the depreciation amount would be constant for the following years? How many years would he be able to depreciate this item? What about the depreciation for the new food truck and the additional equipment? What could he depreciate? In addition, since Diego's business was negatively impacted during the pandemic, he operated with a loss in 2020 and 2021, and he wanted to know if he could reflect some of those losses on the 2022 tax return.

Therefore, Diego's real question was what the rate of return for his business was and how it was affected by taxes and depreciation. Like any businessman he wanted to maximize profits and to reduce his tax burden so he could keep as much money possible for him for future investment.

Reading Activity: Responding to Diego Tax Story
1. What is the role of depreciation?
2. What is the relationship between depreciation and taxation?
3. What kind of items can be depreciated?
4. What is the meaning of bonus depreciation?

Learning Outcomes

After covering this chapter, the student should be able to do the following:

1. Learn what depreciation is and what items can be depreciated.
2. Learn the concept of depreciation recapture.
3. Learn classic and current depreciation methods.
4. Learn how to incorporate depreciation in income tax calculation for businesses.
5. Learn how after-tax calculation affects the IRR.
6. Learn the role of probability and historical data in investment projects.

Introduction: Depreciation and Taxes

On November 13th, 1789, Benjamin Franklin wrote in a letter to Jean-Baptiste Leroy (a French physicist and a major contributor to the *Encyclopédie*), one his most remembered phrases: "Our new Constitution is now established and has an appearance that promises permanency; but in this world nothing can be said to be certain, except death and taxes."

Taxes are needed to support the state and in a sense our way of life. To many taxes are a necessary evil as they have a negative connotation. Every time we see a pay slip, we see how much was subtracted from it, and we rarely consider that through taxes the government is able to provide the basic services that make society work such as basic infrastructure, safety, and public education. Moreover, there are some public services that are not as evident but that provide valuable service to society. For example, thanks to the EPA we enjoy a healthier environment, or thanks to the FDA we can obtain safe medications. As an advocate of economic freedom and I consider that the role of the government should be the one of a referee who acts as regulator that promotes competence and a free market economy.

In this chapter we concentrate on an introduction of the fundamentals of federal income taxes. Federal income taxes are not optional, as they imposed on businesses and individuals by the government through several complex tax laws. Unfortunately, these laws are complex, are modified yearly and have multiple exceptions.

Since taxes have to be paid, they reduce the company's profit and can be considered as another expenditure, and consequently should be included in the business' financial analysis. In addition, depending in the company's location and type of business, income taxes may be collected not only from the federal government but also from state and even local governments. On the other hand, the role of the government is bifold: when businesses generate a profit the government's role is to collect income taxes. However, when businesses have a loss, the government may share the losses.

Because capital investment promotes economic activity and generates wealth, most countries and states promote investment. One way to promote capital investment is by allowing investors to reduce their tax burden by recognizing that a company's capital investments lower their value due to use or obsolescence. For example, as described in the reading, Diego invested in a new food truck, allowing him to depreciate a yearly amount that will lower his tax payment which in turn may help him renew his equipment and continue operations.

Taxable Income and Depreciation

As described by Eliyahu M. Goldratt, the goal of a business is to make money. From their business operations companies generate what is known as revenue. The revenue is then reduced by subtracting all expenses including operation, marketing, financial, ...etc. this amount is known as profit.

The profit is also known as taxable income which is the based considered to determine the amount of taxes that will be paid. Depreciation is then one of the financial expenses that a company must consider in order to determining their taxable income. The role of depreciation is to recognize the decrease in value of an asset due to deterioration or obsolescence. Deterioration occurs when an asset wears out and no longer provides the service provided by new equipment. Obsolescence occurs when new technology appears and replaces the current technology by providing a better service.

From an accounting perspective, depreciation not only comes from a decline in value due to the loss of value to its owner or to the market, but also as a result of a systematic allocation of the cost of an asset divided by its depreciable life. This last perspective is the base for depreciation calculation for tax purposes.

What Can Be Depreciated

There are two different types of property tangible and intangible property. Diego could acquire a new food truck, an office building, or office furniture, and these investments are considered tangible property, as they can be seen, touched, and felt. He can also register his company brand, including its logo, which is intangible property.

Tangible property is divided in real property and personal property. Real property is related to land, buildings, and assets that can be used to build upon, constructed on, or attached to land. On the other hand, some examples of personal property are machinery, equipment, furniture, office equipment, and vehicles.

According to the IRS, businesses cannot depreciate operation expenses; these cash outflows can occur recurrently such as salaries, do not lose gradual value over time, and are used for the main operation of the business over a short period of time (less than one year).

The IRS allows businesses to depreciate tangible property given the following circumstances: the company owns the property, the property is used for business purposes to generate an income, the life of the is more than one year, and the property may lose value gradually over time. Tangible property may not be depreciable when it leased, or not owned by the company, or it is not used for the purpose of generating an income, such as in the case of personal use. In addition, neither land, nor inventory property is depreciable, even when it is in containers and is considered inventory.

Depreciation definition

There are some definitions that have to considered:

- Asset cost (P), the value of an item at the time of purchase. For example, consider that a furniture store buys a new delivery truck for $30,000 in July of 2020. As the truck is used, its value will decrease.
- Book value (BV), updated value for an item for a period posterior to the time of purchase. Using the previous example, assume that the updated book value for the delivery truck after three years of use is $25,000, which is different than the initial purchasing cost.

Equation 8.1Book value

$$\text{Book Value} = (\text{P} - \text{Depreciation charges to date})$$

Market value, the value of an item provided by the market. Continuing with the previous example, a car dealer or an insurance company could value the delivery truck at $20,000 after three years of use, which is different than the starting and updated book values.

Depreciable life is the number of periods in which an asset can be depreciated. This number of periods is determined by tax law. For the delivery truck example, assume that it can be depreciated in seven years. This does not mean that the truck is worthless after seven years, it only means that we have seven years to allocate the loss of value from the truck for tax purposes.

Depreciation Methods

As described, from an accounting perspective depreciation is determined by providing a systematic allocation of an asset's cost over its depreciable life. The following are the 6 methods for calculating depreciation.

1. Straight line (SL)
2. Sum of years digits (SOYD)
3. Double declining balance (DDB)
4. Declining balance (DB) with conversion to straight line SL
5. Unit of production (UOP)
6. Modified accelerated cost recovery system (MACRS)
7. Bonus depreciation

Straight Line Depreciation (SL)

This is the simplest of all methods and is also maybe the most widely known. This method calculates a constant depreciation amount for all periods of the depreciable life.

Equation 8.2 Straight line depreciation

$$\text{SL depreciation} = \frac{(P - S)}{N}$$

Sum of Years Digits Depreciation (SOYD)

In contrast to the SL method, this depreciation method calculates a different depreciation value for each period and allows a more aggressive depreciation in comparison to the SL method by allowing a higher depreciation value to the first periods. This method requires two calculations, first it calculates the SOYD which is then applied to a depreciation equation.

Equation 8.3 Sum of years digits (SOYD)

$$\text{SOYD} = \frac{N}{2}(N + 1)$$

Equation 8.4 SOYD Depreciation charge for any year.

$R_L = Remaninig\ useful\ life\ at\ beginning\ of\ year$

$$\text{SOYD depreciation} = \frac{R_L}{\text{SOYD}}(P - S)$$

Declining Balance Depreciation (DDB)

Like the previous method, DDB calculates a different depreciation value for each period. This method applies a constant depreciation rate to the asset's book value. The Tax Reform Act of 1986 specified two rates of 150% and 200% of the straight-line rate, since 200% is twice the rate of the straight-line it is called double declining balance depreciation.

Equation 8.5. Double declining balance

$$\text{DDB Depreciation} = \frac{(2)}{\text{N}}(\text{P})$$

Example 8.1 performs the depreciation calculation of these three methods.

Example 8.1

A company invested $50,000 in a new computer server with a depreciable life of five years. The computer server has a salvage value of $3,888 after five years. Use the SL, SOYD and DDB methods to determine the depreciation values for the five years.

The depreciation variables are

P = $50,000   S= $3,888   N= 5 years

a) Calculating SL Depreciation

Equation 8.2 Straight line depreciation

$$\text{SL dDepreciation} = \frac{(\text{P} - \text{S})}{\text{N}}$$

$$\text{SL depreciation} = \frac{(\$50{,}000 - \$3{,}888)}{5} = \$9{,}222.40$$

Table 8.1. Straight Line Depreciation

Year	Book Value (BV)	Depreciation for the Year	Book Value after Depreciation
0	$50,000.00		
1	$50,000.00	$9,222.40	$40,777.60
2	$40,777.60	$9,222.40	$31,555.20
3	$31,555.20	$9,222.40	$22,332.80
4	$22,332.80	$9,222.40	$13,110.40
5	$13,110.40	$9,222.40	$3,888.00

Subtotal	$46,112.00

Example 8.1 (Continued)

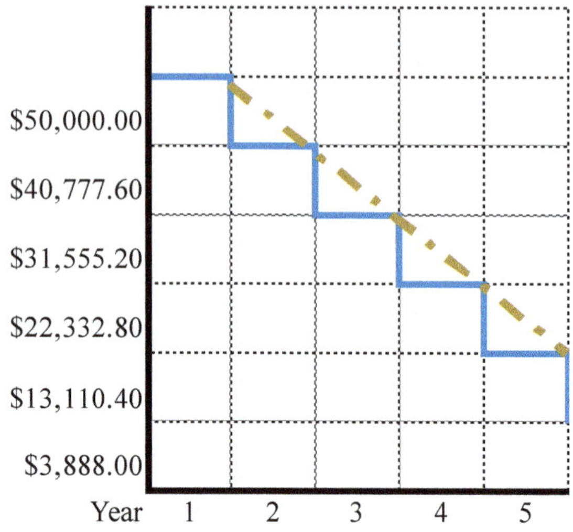

Fig. 8.1. Straight Line Depreciation.

b) Calculating SOYD Depreciation

Equation 8.3 Sum of years digits (SOYD)

$$\text{SOYD} = \frac{N}{2}(N+1) = \frac{5}{2}(5+1) = 15$$

Equation 8.4 SOYD depreciation charge for any year.

$R_L = Remaninig\ useful\ life\ at\ beginning\ of\ year$

$$\text{SOYD depreciation} = \frac{R_L}{\text{SOYD}}(P-S)$$

For year 1, $R_L = 5$:

$$\text{SOYD depreciation} = \frac{5}{15}(\$50{,}000 - \$3{,}888) = \$34{,}629.33$$

For year 2, $R_L = 4$:

$$\text{SOYD depreciation} = \frac{4}{15}(\$50{,}000 - \$3{,}888) = \$12{,}296.53$$

Example 8.1 (Continued)

Table 8.2 SOYD Depreciation

Year	Book Value (BV)	Depreciation for the year	Book Value after depreciation
0	$50,000.00		
1	$50,000.00	$15,370.67	$34,629.33
2	$34,629.33	$12,296.53	$22,332.80
3	$22,332.80	$9,222.40	$13,110.40
4	$13,110.40	$6,148.27	$6,962.13
5	$6,962.13	$3,074.13	$3,888.00

Subtotal	$46,112.00

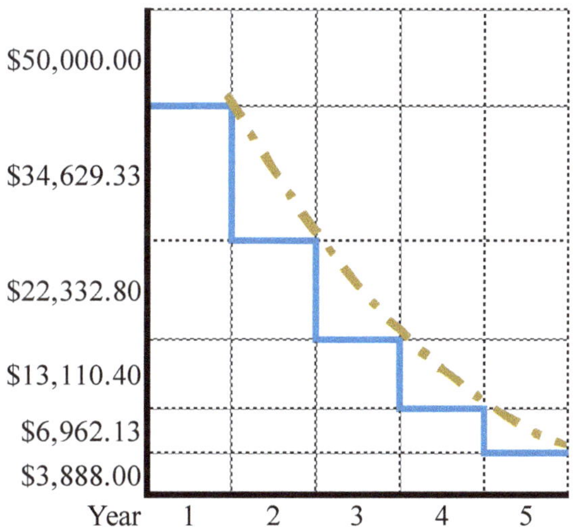

Fig. 8.2. SOYD Depreciation.

Example 8.1 (Continued)

c) Calculating declining balance depreciation

Equation 8.5. Declining balance depreciation

$$\text{DDB Depreciation} = \frac{(2)}{N}(P)$$

For year 1

$$\text{DDB depreciation} = \frac{(2)}{5}(\$50{,}000) = \$20{,}000$$

For year 2

$$\text{DDB depreciation} = \frac{(2)}{5}(\$30{,}000) = \$12{,}000$$

Table 8.3 Declining Balance Depreciation

Year	Book Value (BV)	Depreciation for the Year	Book Value after Depreciation
0	$50,000.00		$0.00
1	$50,000.00	$20,000.00	$30,000.00
2	$30,000.00	$12,000.00	$18,000.00
3	$18,000.00	$7,200.00	$10,800.00
4	$10,800.00	$4,320.00	$6,480.00
5	$6,480.00	$2,592.00	$3,888.00
	subtotal	$46,112.00	

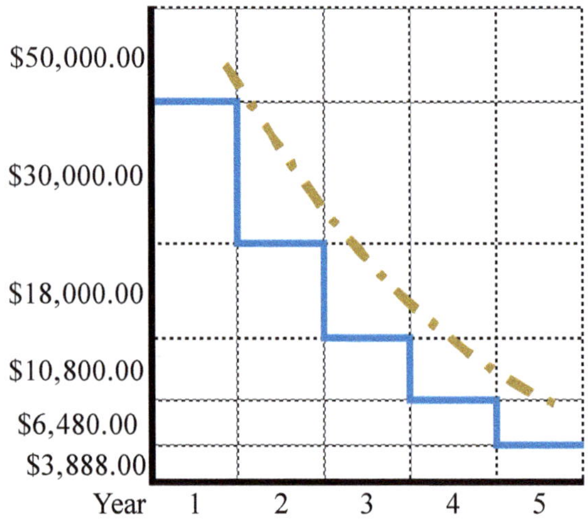

Fig. 8.3. Declining Balance Depreciation.

Declining Balance Depreciation in any year

From Example 8.1 we learned the first three depreciation methods and we may observe how the depreciation for the initial periods is higher for the DDB than SOYD and for SOYD than for SL.

The following describes the formulation for the Declining Balance for any year. Considering Equation 8.5 as the depreciation for year 1:

Equation 8.5. Declining balance depreciation for year 1

$$\text{DDB depreciation} = \frac{(2)}{\text{N}}(\text{P}) = \frac{(2\text{P})}{\text{N}}$$

We the calculate the depreciation for year 2 by updating the value of P by $\left(P - \frac{2P}{N}\right)$:

$$\text{DDB depreciation for year 2} = \frac{(2)}{\text{N}}\left(P - \frac{2P}{N}\right) = \frac{2P}{N}\left(1 - \frac{2}{N}\right)$$

For year 3 we follow a similar approach 3 by updating the value of P by $\left(P - \frac{2P}{N}\right)$:

$$\text{DDB depreciation for year 3} = \frac{(2)}{\text{N}}\left(P - \frac{2P}{N} - \frac{2P}{N}\left(1 - \frac{2}{N}\right)\right) = \frac{2P}{N}\left(1 - \frac{2}{N}\right)^2$$

The depreciation for year *n* is calculated in Equation 8.6:

Equation 8.6 Declining balance depreciation for year n

$$DDB\ Depreciation\ for\ year\ n = \frac{2P}{N}\left(1 - \frac{2}{N}\right)^{n-1}$$

In the event that of declining balance depreciation is set to 150%, the number 2 in Equation 8.6 is replaced by a value of 1.5. The total declining balance depreciation at the end of *n* years is expressed as

Equation 8.7 Total declining balance depreciation after "n" years

$$DDB\ Depreciation\ for\ year\ n = P\left[1 - \left(1 - \frac{2}{N}\right)^n\right]$$

Therefore, the Book Value for an asset (P) minus the total depreciation at the end of *n* years is

Equation 8.8 Book value for declining balance depreciation at the end of n years

$$Book\ Value\ for\ year\ n = P - P\left[1 - \left(1 - \frac{2}{N}\right)^n\right] = P\left(1 - \frac{2}{N}\right)^n$$

Exercise

Determine the total depreciation values using SL, SOYD, and DDB for an item with a value of $10,000 and a salvage value of $1,500 after five years.

Answer: SL $ 5,000 SOYD $ 5,000 DDB $9,222.4

The results provided from the previous exercise show that the DDB has a total depreciation value different than the value calculated for SL and SOYD this is because the salvage value provided the SL and SOYD calculations was too low.

On the other hand, the previous example also shows that choosing DDB generates higher depreciation values for the first years in comparison to SOYD and SL.

Influence of Salvage Value on Declining Balance Depreciation

After determining the depreciation value, we could obtain three different situations, these are shown in Fig. 8.4. The first one, A, shown with a green dotted line occurs when the depreciation value is lower than the salvage value. The second case, B, shown with a red line follows when the depreciation value is equal to the salvage value and the third one, C, shown with a black line happens when the depreciation value is higher than the salvage value.

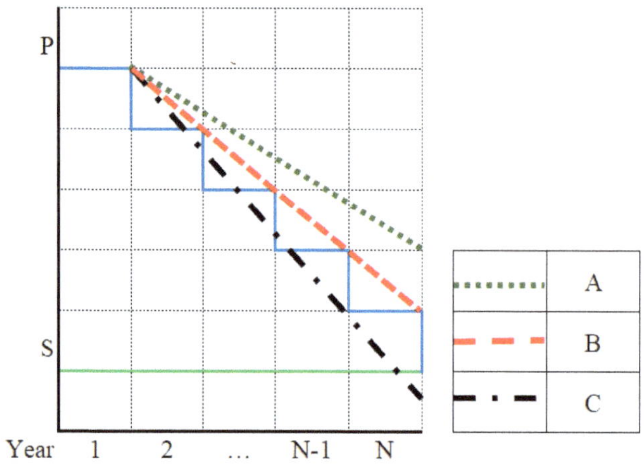

Fig. 8.4. Effect of salvage value on DBD.

To understand these scenarios let us start with B; since the depreciation value is equal to the salvage value there is no problem as there is no difference. C shows that the company depreciated for an amount higher than item's salvage value; this is not allowed by the IRS, as the depreciation should end when the difference between the depreciation and the salvage value is zero.

A shows that the depreciation value is lower than the item's salvage value. This situation will always occur if the salvage value is equal to zero. For this situation there are two possible solutions:

a) Continue using the asset until and keep depreciating until the depreciation reaches the salvage value, which could be zero.

b) Transfer from a DDB to a composite depreciation. The IRS allows companies to change from the declining balance method to straight depreciation SL at any time. Therefore, in the last period of the remaining useful life a company can account for the remaining book value–salvage value. This brings us to the fourth depreciation method.

Declining Balance Depreciation (DDB) with Conversion to Straight-Line Depreciation

As shown in Example 8.1 the depreciation amount for year 1 using DDB was higher than the amount for SL. Generally speaking, companies want to depreciate the book value of an asset as quickly as possible due to the value of time and money, as described in Chapter 2. Therefore, companies switch to SL depreciation whenever the value of SL depreciation is higher than the depreciation value from DDB or other depreciation methods. However, we should note that once the depreciation method is changed to SL it cannot be switched back to the previous method. Example 8.2 shows an application of this situation.

Example 8.2

A construction company bought a new tower crane and invested $540,000 with a depreciable life of seven years. The tower crane has a salvage value of $27,000 after seven years of use. Perform SL and DDB depreciation for this item and determine when the company should switch to SL depreciation.

$P = \$540{,}000 \quad S = \$27{,}000 \quad n = 7\ years$

Table 8.4 Declining balance depreciation Example 8.2

	Declining Balance Depreciation (DBD)			Straight Line (SL)	Decision
Year	Book Value (BV)	Depreciation for the Year	Book Value after Depreciation		
0	$540,000.0		$0.0		
1	$540,000.0	$154,285.7	$385,714.3		
2	$385,714.3	$110,204.1	$275,510.2	$59,785.7	Maintain DBD
3	$275,510.2	$78,717.2	$196,793.0	$49,702.0	Maintain DBD
4	$196,793.0	$56,226.6	$140,566.4	$42,448.3	Maintain DBD
5	$140,566.4	$40,161.8	$100,404.6	$37,855.5	Maintain DBD
6	$100,404.6	$28,687.0	$71,717.6	$36,702.3	Convert to SL
7	$71,717.6	$20,490.7	$51,226.8		
	subtotal	$488,773.2			

The DBD computation for year 1 is as follows

$$D1 = \frac{2}{7} * \$540{,}000 = \$154{,}285.7$$

The DBD computation for year 2 is as follows

$$D2 = \frac{2}{7} * \$385{,}714.3 = \$110{,}204.1$$

Example 8.2 (Continued)

The SL computation for year 2 is as follows:

$$D2 = \frac{(\$385{,}714.3 - \$27{,}000)}{7 - 1} = \$59{,}785.7$$

Since the values for DBD are higher than the values for SL for years 2–5 we maintain DBD depreciation for these years and convert to SL depreciation for year 6 with a value of $36,702.3 and $35,015.27 for year 7. It cannot be the same value for both years because the depreciation would be higher than the value of the asset.

Unit-of-Production Depreciation (UPD)

There are some situations in which the depreciation of a particular asset is more closely related to the wear and tear from the number of units produced, than the age of the asset. As a result, this approach often provides a higher depreciation amount for the years the asset is under heavy operation.

As shown in Equation 8.9, the depreciation is calculated by subtracting the salvage value (S) from the asset's original cost (P) and then multiplying this amount by the number of units produced. This subtotal is then divided by the expected number of units that the asset can manufacture/produce during its useful life.

Equation 8.9. Unit of production depreciation

$$UPD = \frac{(P - S) * Units}{Production\ Capability}$$

Example 8.3

A car manufacturing company uses certain press to mold a part used in aerospace. The part is mainly used in older planes and therefore the demand is variable. The press has a cost of $1,000,000 with a salvage value of $50,000. Assume that the press has a production capability of 25,000 units and that the number of parts required is as shown and determine the depreciation amount per year using UPD.

Year	Demand (units)
1	3000
2	6000
3	12000
4	3000
5	1000

Example 8.3 (Continued)

$$UPD = \frac{(P - S) * Units}{Production\ Capability}$$

For year 1

$$UPD = \frac{(\$1{,}000{,}000 - \$50{,}000) * 3{,}000}{25{,}000} = \$114{,}000$$

Table 8.5 Yearly Depreciation values using UPD

Year	Demand (Units)	Depreciation
1	3000	$114,000
2	6000	$228,000
3	12000	$456,000
4	3000	$114,000
5	1000	$38,000

Modified Accelerated Cost Recovery System Depreciation (MARCS)

The modified accelerated cost recovery system depreciation (MARCS) is the newest depreciation in the tax law. This method was introduced in the Tax Reform Act of 1986. The MARCS method is the most common procedure for calculating depreciation.

To simplify depreciation calculations using MARCS the salvage value is assumed to be equal to zero. In addition, the IRS classified assets by class and assigned a service life described as asset depreciation range (ADR). ADR considers a lower life limit, a midpoint life limit, and an upper life limit, where the midpoint life limit is lower than the average useful lives.

The MARCS method is divided into two main categories:

a) General depreciation system (GDS). This is the main method of depreciation for tangible property used for deductions income tax calculations for economic activities generated in the United States. This tax deduction system allows higher depreciation amounts in the early years and lower depreciation amounts in the later years.

b) Alternative depreciation system (ADS). ADS is used for income tax calculations for tangible property used for economic activities generated outside the United States. Generally, the GDS allows higher depreciation in the early years than ADS as GDS considers shorter service lives. In addition, ADS may be elected for an asset, but it cannot be switch back to GDS.

Equation 8.10 shows that the depreciation calculation (d_t) for a given year (t) is determined by multiplying the asset's cost (P) by the depreciation rate allowed (r_t) according to the asset's life.

Equation 8.10 MARCS depreciation formula

$$D_t = P * r_t$$

To calculate the MARCS depreciation, we follow the next steps:

1. Establish asset's cost basis (P). This cost basis may include the asset's cost, and additional fees and charges.
2. Learn the property class and recovery period as described in Table 8.6. Use GDS unless explicitly required by law to use ADS.
3. Determine asset's placed-in-service date. This is relevant as the depreciation amount is proportional to the in-service date.

Table 8.6. MARCS 3-, 5-, 7-, 10-, 15-, and 20-Year Property Half-Year Convention. [1]

Year	3-year	5-year	7-year	10-year	15-year	20-year
1	33.33%	20.00%	14.29%	10.00%	5.00%	3.750%
2	44.45%	32.00%	24.49%	18.00%	9.50%	7.219%
3	14.81%	19.20%	17.49%	14.40%	8.55%	6.677%
4	7.41%	11.52%	12.49%	11.52%	7.70%	6.177%
5		11.52%	8.93%	9.22%	6.93%	5.713%
6		5.76%	8.92%	7.37%	6.23%	5.285%
7			8.93%	6.55%	5.90%	4.888%
8			4.46%	6.55%	5.90%	4.522%
9				6.56%	5.91%	4.462%
10				6.55%	5.90%	4.461%
11				3.28%	5.91%	4.462%
12					5.90%	4.461%
13					5.91%	4.462%
14					5.90%	4.461%
15					5.91%	4.462%
16					2.95%	4.461%
17						4.462%
18						4.461%
19						4.462%
20						4.461%
21						2.231%

The depreciation percentages shown in Table 8.6 do not follow the life duration described in the column heading and consider an extra year. For example, the 7-year column includes a depreciation rate for year 8. This is because asset depreciation follows the half-of-year convention in which assets are assumed to be placed in service at half of the year and also because assets are assumed to be fully depreciated with a salvage value of zero at the end of the depreciable life.

[1] https://www.irs.gov/publications/p946#en_US_2020_publink100068700 (accessed: 12/12/2022)

As described, assets are classified by property classes. Some examples of the different asset classes are described in table 8.7.

Table 8.7. Examples of table of class lives [2]

Property Class	Examples of table of class lives
3-year property	Personal Property (all property except real state) Manufacture of Food and Beverages - Special Handling Devices Manufactue of Finished Plastic Products - Special Tools Manufacture of Glass Products - Special Tools Manufacture of Fabricated Metal Products - Special Tools
5-year property	Data Handling Equipment; except computers Airplanes (aiframes and engines) Automobiles, Taxis Buses, General Prupose Trucks Trailers and Trailer-Mounterd Containers
7-year property	Office Furniture, Fixtures and Equipment Railroad Cars and Locomotives Citton Ginning Assets Assets used in mining Manufacture of Other Food and Kindred Products
10-year property	Vessels, Barges, Tugs, and Similar Water Transportation Equipment Single prupose agricultural or horticultural structures Petroleum Refining Manufacture of Grain and Grain Mill Products Manufacture of Sugar and Sugar Products
15-year property	Land Improvements Industrial Steam and Electric Generation and/or Distribution Systems Manufacture of Cements Railroad Wharves and Docks Water Transportation Pipeline Transportation
20-year property	Railroad Structures and Similar Improvements Telephone Communication Railroad Hydraulic Generating Equipment TOCSC - Central Office Control Equipment+AJ70:AK75

[2] https://www.irs.gov/publications/p946#en_US_2020_publink1000107772 (accessed: 12/12/2022)

Example 8.4

A healthcare company uses cars to visits their customers. Determine the depreciation schedule for a vehicle with a cost of $45,000 using MARCS GDS.

The depreciation values per year are calculated by applying equation 8.10

$$D_t = P * r_t$$

Multiply asset value *P* by the MARCS rate as shown in the following table.

Table 8.8 Example 7 Depreciation table per period and cumulative using MARCS GDS

Year	Book Value (BV)	MARCS GDS Rate	Yearly Depreciation	Cumulative Depreciation	Current Book Value
0	$45,000.0				
1	$45,000.0	20.00%	$9,000.00	$9,000.00	$36,000.00
2	$45,000.0	32.00%	$14,400.00	$23,400.00	$21,600.00
3	$45,000.0	19.20%	$8,640.00	$32,040.00	$12,960.00
4	$45,000.0	11.52%	$5,184.00	$37,224.00	$7,776.00
5	$45,000.0	11.52%	$5,184.00	$42,408.00	$2,592.00
6	$45,000.0	5.76%	$2,592.00	$45,000.00	$0.00

Depreciation Recapture and Capital Gain

When an asset is acquired, depreciated, and then sold it must be determined if the selling price was greater than the asset's original book value (BV); if the selling price is greater than the current BV, or if the selling price is below the current book value. Answering this question is relevant as it affects the income tax amount. To clarify, consider a scenario in which a car rental company sells a car after three years of use. Based on the original BV, the current BV, and the selling price of the item we obtain the following three situations:

a) The company would get a **capital gain** if the selling price was **greater** than the original BV. In such a case, the capital gain is calculated by subtracting the original BV from the selling price.

b) The company would obtain a **depreciation recapture** if the selling price was higher than the current BV. The depreciation recapture is the difference between the selling price and the current BV

c) The company would have a **loss** if the selling price was **lower** than the current BV, and the loss is calculated as the difference between the current BV and the selling price.

This situation is described in Example 8.5.

Example 8.5

Bananas car rental is a car rental company, that acquired an economy car for $17,000. Assume that after applying a MARCS GDS depreciation for three years the company decided to sell this car. Determine the car's BV in year 3 and the capital gain, depreciation recapture, and loss if the car was sold for (a) $10,000, (b)$1,500 and (c)$20,000.

Table 8.9 Depreciation Table for Year 3 Example 8.5

Year	Book Value (BV)	MARCS GDS Rate	Yearly depreciation	Cumulative Depreciation	Current Book Value
0	$17,000.00				
1	$17,000.00	20.00%	$3,400.00	$3,400.00	$13,600.00
2	$17,000.00	32.00%	$5,440.00	$8,840.00	$8,160.00
3	$17,000.00	19.20%	$3,264.00	$12,104.00	$4,896.00

Therefore,

Original BV =$17,000

Current BV in year 3 =$4,896

a) Sell price =$10,000

Capital gain = $0; the selling price is **lower** than the original BV.

Depreciation recapture $= \$10{,}000 - \$4{,}896 = \$5{,}104.00$

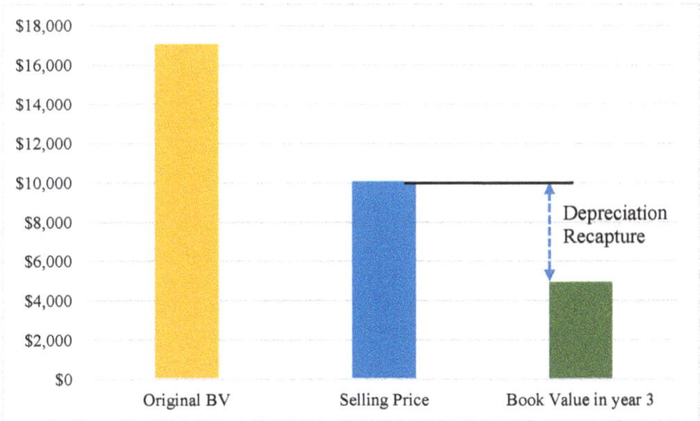

Fig. 8.5. Capital gain loss is higher than sell price.

Example 8.5 (Continued)

b) Sell price =$1,500

Capital gain = $0; the selling price is **lower** than the original BV.

Depreciation recapture = $0; the selling price is **lower** than the BV in year 3.

Loss = $4,896 – $1,500 = $3,396.00

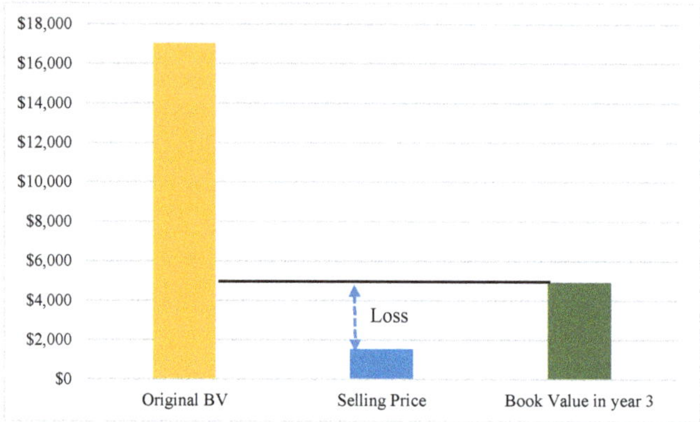

Fig. 8.6. Loss sell price is lower than BV.

c) Sell price =$20,000

Capital gain = $20,000 – $17,000 = $3,000.00; the selling price is **higher** than the original BV.

Depreciation recapture = $17,000 – $4,896.00 = $12,104.00; the selling price is **higher** than the BV in year 3.

Loss = $0

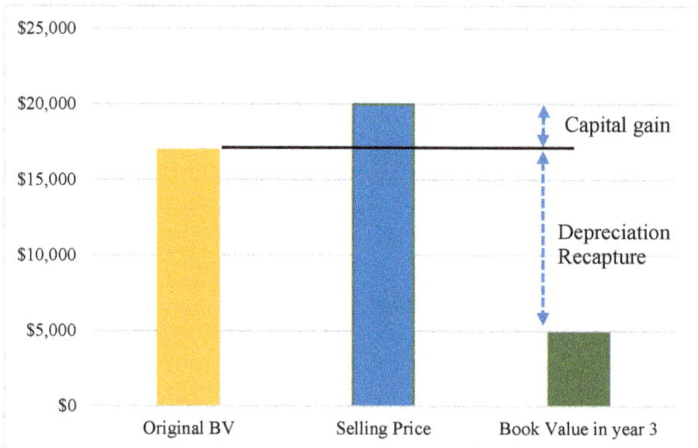

Fig. 8.7. Capital gain and depreciation recapture.

Depletion

Depletion occurs when the source of a natural resource is finished after being exploited. For example, a gold mine is depleted once it runs out of gold. Since depletion is related to natural resources it can occur in several industries such as oil and gas, mining, and timber. As the natural resource is exploited the amount of the resource left at the source is decreased. Therefore, depletion means that a company is consuming their capital investment, and for tax purposes this consumption is considered a capital loss.

There are two methods to calculate depletion, cost depletion and percentage depletion, and the largest value obtained from these methods is the one considered for tax calculations as depletion amount. Cost depletion follows a similar calculation as UPD. Here, we consider the property cost, the estimated value of recoverable units, and the salvage cost (if there is any), as observed in Example 8.6.

Example 8.6

A high-end restaurant chain builds their furniture from a distinctive type of wood. In order to have enough wood for current and future developments they bought a lot for $950,000; they estimated $250,000 for the land, and $700,000 for the timber, with an estimate volume of 4 million board-feet of standing timber.

Assume that the company cut 50,000 board-fleet of standing timber during the first year. Determine the depletion allowance for the year using cost depletion.

Allowance per 1000 board-ft

$$Allowance\ per\ board-ft = \frac{\$950{,}000 - \$250{,}000}{4{,}000 * 1000\ board-ft}$$

$$Allowance\ per\ board-ft = \frac{\$700{,}000}{4{,}000{,}000} = \$0.175\ board-ft$$

$$Allowance\ per\ 1000\ board-ft = 1{,}000 * \$0.175\ board-ft = \$175$$

Therefore, the allowance for the 1st year

$$50{,}000\ board-ft * \ \$0.175\ board-ft \ = \$8{,}750$$

Percentage Depletion

This depletion method considers a percentage of the gross income from the property for the given year and the type of mineral that is extracted. The percentages are described in Table 8.10. This depletion calculation method is entirely different to depreciation because it considers the gross income and not the initial cost of the asset.

On the other hand, because percentage depletion is calculated by a percentage of the gross income, the depletion calculation can be higher than the original value of the asset. For example, consider a gold mine that, due to a new technical advancement, can now extract a high amount of gold and that was acquired for a low price several years ago. For such situations the percentage depletion allowance in any year is limited to not more than 50% of the taxable income from the property, calculated without the depletion deduction, as shown in Example 8.7.

Table 8.10 Percentage Depletion Allowances 26 U.S. Code § 613 - Percentage depletion

Deposits	Rate
Sulphur and uranium; and deposits in the United States of anorthosite, clay, laterite, and nephelite syenite, asbestos, bauxite, celestite, chromite, corundum, fluorspar, graphite, ilmenite, kyanite, mica, olivine, quartz crystals (radio grade), rutile, block steatite talc, and zircon, and ores of the following metals: antimony, beryllium, bismuth, cadmium, cobalt, columbium, lead, lithium, manganese, mercury, molybdenum, nickel, platinum and platinum group metals, tantalum, thorium, tin, titanium, tungsten, vanadium, and zinc.	22%
Corporate deduction for iron ore and coal.	20%
Geothermal deposits, gold, silver, copper, iron ore, and certain oil shale, if from deposits in the U.S.	15%
Borax, granite, limestone, marble, mollusk shells, potash, slate, soapstone, and carbon dioxide produced from a well.	14%
Natural gas from geopressured brine. Coal, lignite, and sodium chloride.	10%
Clay and shale used or sold for use in making sewer pipe or bricks or used or sold for use as sintered or burned lightweight aggregates.	7.5%

Example 8.7

A small gold mine with a small gross income for the previous year reached $2,500,000. The company had expenses of $450,000. Determine the allowable percentage of depletion deduction.

Considering the rate information from Table 8.10, the depletion percentage is 15%.

The allowable percentage depletion deduction is limited to a maximum of 50% of the taxable income.

Since the taxable income limitation value is higher than the computed percentage depletion, the depletion value is $375,000.

Example 8.7 (Continued)

A small gold mine with a small gross income for the previous year reached $2,500,000. The company had expenses of $450,000. Determine the allowable percentage of depletion deduction.

Computed Percentage Depletion

Gross income	$2,500,000
Depletion Percentage	15%
Calculated depletion	$375,000

Taxable Income Limitation

Gross income	$2,500,000
Expenses (minus)	$450,000
Taxable income	$2,050,000
Deduction limitation	50%
Taxable income limitation	$1,025,000

Taxable Income

In the initial part of this chapter, we learned the basics about depreciation is and how it can be calculated. We also learned that the purpose of depreciation is to consider the asset's wear/obsolescence in order to reduce the amount of taxable income. Income taxes for individuals and businesses follow very different rules. Individuals' income-tax follows a progressive structure, meaning that individuals pay a higher-income tax rate as they earn more money. On the other hand, for businesses the rules changed in 2017 when the Tax Cuts and Jobs Act (P.L. 115-97) replaced the previous tax structure with a flat 21% corporate tax rate.

Individuals' taxable income

To understand individuals taxable income, it is important to consider the following concepts adjusted gross income (AGI) and modified adjusted gross income (MAGI). IRS uses the AGI and MAGI values to determine tax liability and if a person may qualify for certain tax benefits.

AGI is calculated by first calculating gross income by adding all sources of income such as wages (salary), income from investments, capital gains, dividends, unemployment compensation, and withdrawals from a 401(k) retirement fund. Once the gross income is calculated we obtain the AGI by subtracting the allowable adjustments such as tax-exempt interest income, and other tax deductions such as alimony payments (for divorces before 2019), IRA deductions, educator expenses, contributions to health savings accounts (HSA), charitable contributions, home mortgage interest, property tax, and student loan interest.

MAGI considers the AGI value and takes back certain adjustments such as tuition-related costs or deductions, rental property losses, and student loan interests, among other adjustments. As gross income increases individuals may apply fewer/lower deductions, for example on IRA account contributions.

Therefore, MAGI and AGI consider different tax deductions and applicable credits, which affect an *individual's* tax return. In the event that a person does not have tax-exempt interest income nor tax deductions, the value of their MAGI is equal to the value of their AGI. On the other hand, if a person receives nontaxable Social Security benefits, tax-exempt interest, or untaxed foreign income they have to add these benefits to their AGI to determine their MAGI.

Business Taxable Income

Gross income comes from the businesses comes from the business's main economic activity, plus other income such as investments, capital gains, dividends and earned interest, and from selling assets, among many others. The taxable income is the difference between gross income minus expenditures. Expenditures include depreciation charges and depletion charges but do not consider capital expenditures.

In addition, business expenditures can be classified in three main categories:

1. Expenditures related to depreciable assets for assets with a life greater than one year
2. Expenditures related to non-depreciable assets that have a life lower than 1 year or that cannot be depreciated such as land
3. All other business expenditures, including those that may be necessary such as utilities or not be necessary, such as donations. Other types of business-related and unrelated expenditures as described in Table 1.1, that cannot be depreciated.

Example 8.8

IMG 8.2

Vega's bookstore is a local business that had sales for $300,000 in year 1, $350,000 in year 2, and $400,000 in year 3. In order to improve their brand recognition, Vega's acquired a store mascot figure of "Kunito the Reader Racoon." The 5-feet figure had a cost of $60,000 and is part of the business personal property.

The bookstore had yearly expenses for $120,000 in year 1, $130,000 in year 2 and $140,000 in year 3. Determine and compare the after-tax income for the first three years of operation considering the following situations:

(a) the figure can be depreciated at the end of each year following MARCS GDS

(b) the figure can be depreciated at the end of each year following MARCS GDS with a 10% bonus depreciation in year 1

(c) the figure could be fully depreciated at the end of the first year

Example 8.8 (Continued)

a) Assuming a three-year personal property depreciation we obtain yearly depreciation:

Table 8.11 Depreciation values assuming a three-year personal property depreciation Example 8.8

Year	Book Value (BV)	MARCS GDS Rate	Yearly depreciation	Cumulative Depreciation	Current Book Value
1	$60,000.00	33.33%	$19,998.00	$19,998.00	$40,002.00
2	$40,002.00	44.45%	$26,670.00	$46,668.00	$13,332.00
3	$13,332.00	14.81%	$8,886.00	$55,554.00	$4,446.00
4	$4,446.00	7.41%	$4,446.00	$60,000.00	$0.00

We calculate the after-tax income and tax liability for the different MARCS GDS:

Table 8.12 Taxable income for Table 8.11

	Year 1	Year 2	Year 3
Gross income	$300,000	$350,000	$400,000
Figure Depreciation	$19,998	$26,670	$8,886
All other expenditures	$120,000	$130,000	$140,000
Taxable income	$160,002	$193,330	$251,114
Tax liability 21.0%	$41,825	$50,536	$65,641
After tax income	$118,177	$142,794	$185,473

(b) Assuming a three-year personal property depreciation we obtain yearly depreciation plus a 10% bonus depreciation in year 1year.

Table 8.13 Depreciation values assuming a three-year personal property depreciation plus a 10% bonus depreciation Example 8.8

Year	Book Value (BV)	MARCS GDS Rate	Yearly depreciation	Bonus Depreciation	Cumulative Depreciation	Current Book Value
0	$60,000.00					
1	$60,000.00	33.33%	$19,998.00	$6,000.00	$25,998.00	$34,002.00
2	$34,002.00	44.45%	$26,670.00		$52,668.00	$7,332.00
3	$7,332.00	12.22%	$7,332.00		$60,000.00	$0.00

Example 8.8 (Continued)

We calculate the after-tax income for the different MARCS GDS with 10% bonus depreciation:

Table 8.14: Taxable income for Table 8.13

	Year 1	Year 2	Year 3
Gross income	$300,000	$350,000	$400,000
Figure Depreciation	$25,998	$26,670	$7,332
All other expenditures	$120,000	$130,000	$140,000
Taxable income	$154,002	$193,330	$252,668
Tax liability 21.0%	$40,256	$50,536	$66,047
After tax income	$113,746	$142,794	$186,621

c) Assuming a 100% bonus depreciation in year 1 MARCS GDS cannot be calculated as we cannot depreciate a value higher than 100%. Table 8.15 shows the taxable income for this scenario.

Table 8.15 Taxable income for 100% bonus depreciation in year 1

	Year 1	Year 2	Year 3
Gross income	$300,000	$350,000	$400,000
Figure Depreciation	$60,000	$0	$0
All other expenditures	$120,000	$130,000	$140,000
Taxable income	$120,000	$220,000	$260,000
Tax liability 21.0%	$31,368	$57,508	$67,964
After tax income	$88,632	$162,492	$192,036

In this example we observe a difference for after tax income for year 1 which changed from $118,177 to $88,632.

This could be interpreted in two different ways; one perspective can consider the after-tax income of $88,632 as good news because tax liability was low for a $300,000 gross income.

The second perspective may consider that the after-tax income of $88,632 is due to a poor performance, which is an incorrect consideration as the lowering the taxable income allows companies to lower their tax burden.

Combining Federal and State Income Tax

In addition to federal income taxes most individuals and corporations are also required to pay state income tax. Since state taxes are deductible when determining federal income tax, the state income tax is applied to the larger taxable income. Therefore, combining federal and state income tax does not equate to adding tax rates.

The following equation considers the calculation of the combined income tax rate for state and federal income tax.

Equation 8.11 Combined income tax rate from federal and state income tax

$$\text{Combined Inc Tax Rate} = \text{State Inc Tax Rate} + [\text{Federal Inc Tax Rate} * (1 - \text{State Inc Tax Rate})]$$

Example 8.9

Vega's bookstore from Example 8.7 is located in New York. Since the business's sales were lower than for $5,000,000 the state tax rate is 6.5% (otherwise it would be 7.25%). Determine the combined income tax rate and the tax liability for each period, consider the depreciation information described in Table 8.11.

Federal income tax rate = 21%

NY State income tax rate = 6.5%

Using Equation 8.12

$$\text{Combined Inc Tax Rate} = 6.5\% + [21\% * (100\% - 6.5\%)]$$

$$\text{Combined Inc Tax Rate} = 6.5\% + (21\% * 93.5\%) = 6.5\% + 19.635\% = 26.14\%$$

Table 8.16 Combined tax rate and tax liability

	Year 1	Year 2	Year 3
Gross income	$300,000	$350,000	$400,000
Depreciation of the figure	$19,998	$26,670	$8,886
All other expenditures	$120,000	$130,000	$140,000
Taxable income	$160,002	$193,330	$251,114
Tax liability 26.14%	$41,825	$50,536	$65,641
After tax income	$118,177	$142,794	$185,473

Considering Income Taxes for Rate Of Return (ROR) Calculation

In previous chapters we did not consider the taxes liability for the calculation of the rate of return. Unfortunately, this is unrealistic and misleading. Expenses, including depreciation and income taxes must be considered for the correct calculation of the rate of return.

Example 8.10

Vega's bookstore would like to know the ROR; use consider the information provided in the previous examples and assume the store's initial investment was $300,000.

Consider the information from Table 8.12 for before-tax income and after-tax income.

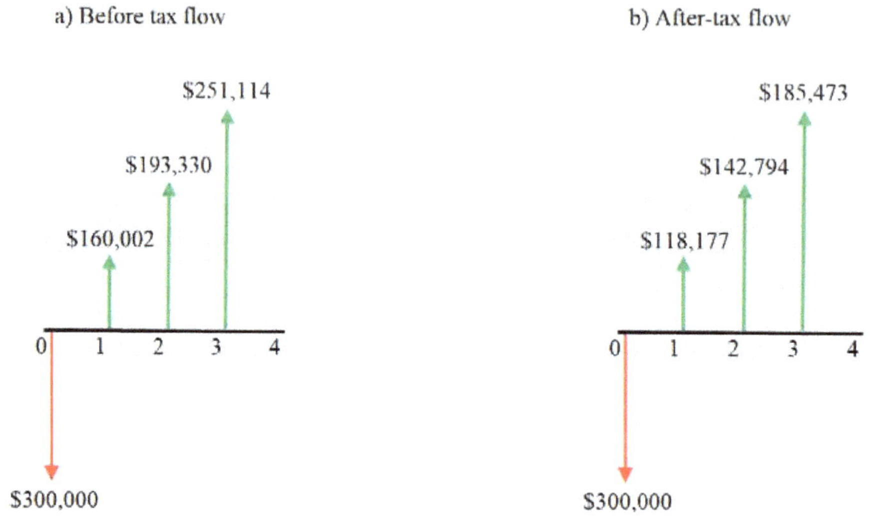

Fig. 8.8 Cash flow for Vega's bookstore a) before tax flow b) after tax flow.

The equation for the ROR for the before-tax flow is as follows:

$$PW = -\$300{,}000 + \$160{,}002\,(P/F, i\%, 1) + \$193{,}330\,(P/F, i\%, 2) + \$251{,}114\,(P/F, i\%, 3)$$

The equation for the ROR for the after-tax flow is as follows:

$$PW = -\$300{,}000 + \$118{,}177\,(P/F, i\%, 1) + \$142{,}794\,(P/F, i\%, 2) + \$185{,}473\,(P/F, i\%, 3)$$

We use Excel to solve these equations

	Before tax	After-tax
Year 0	-$300,000	-$300,000
Year 1	$160,002	$118,177
Year 2	$193,330	$142,794
Year 3	$251,114	$185,473
Rate of return	41.1%	21.0%

We use Excel function to calculate the ROR for both options, the ROR is clearly reduced from the paying taxes. The ROR calculation is described in Chapter 6.

As observed in the ROR from Example 8.10 there are clear differences between before- and after-tax calculations. The subtotal before income tax reached $604,446.00, while the after-tax subtotal reached $446,443.82. This clear difference describes the relevance of considering taxes, licenses, and all other additional expenses that may affect the ROR for an investment.

What Did We Learn in This Chapter?

1. We learned what depreciation is and what assets can be depreciated.
2. We learned different methods to calculate depreciation.
3. We learned how to calculate taxable income.
4. We learned how to determine federal income tax liability for federal tax and for federal and state tax combined.
5. We learned how to determine the rate of return in a project before and after income tax.

Review Questions

Directions: Refer to what you learned in this chapter in order to answer the following questions:

1. Why do companies want to depreciate their assets?
2. Considering depreciation, what are the advantages from buying a two-year-old car instead of a new car?
3. What is the difference between depletion and depreciation?
4. Is the book value constant or variable?
5. How many types of MARCS depreciation are there, and what is the difference between them?
6. How does a bonus depreciation promote investment?
7. The personal income tax has a progressive structure, and the federal income tax uses a flat rate of 21%. What are the advantages of one system versus the other?
8. Why should we use before- and after-tax information when calculating the rate of return?

Chapter 8 Practice Problems

1. A computer company is investing in a new conveyor system to move components in a distribution center. They are planning on spending $3,500,000. Assume that the system has a salvage value of 20% after seven years of use. Determine the depreciation values per year using the following methods:

    a. Straight line depreciation

    b. Sum of years digits depreciation

    c. Declining balance depreciation.

2. Using the information from question 1, determine the depreciation per year using MARCS GDS.

3. A local business acquired a new SUV, (consider it a five-year item). The SUV has a value of $85,000. Calculate the DDB, the SL depreciation, and the MARCS depreciation. See if there is a switch between DDB to SL.

4. A company acquired a three-year item for a value of $150,000. Assume that after applying a MARCS GDS depreciation the company decided to renew the item and sell it after five-year. Assume that the book value at year 5 is $0. Determine the item's capital gain, depreciation recapture, and loss if the item was sold for (a) $50,000, (b)$15,000 and (c)-$20,000 (the company paid to have it removed).

5. Determine the income tax liability for a company with $3,500,000 in sells, $575,000, in expenses and with a depreciation charge of $90,000.

6. A new company with less than 1,000,000 in revenue wants to find the best location for a new manufacturing facility. Determine the combined taxable rate (federal and state income tax) for Texas, Illinois, Florida, and New York, and determine the best location based on tax rate.

7. Kunito Co is an electronics manufacturer. Purchased a large fleet of cars for his sales representatives. The fleet had a cost of $2,000,000. Cars are considered a 5-year class. In their first year of operation, they had a revenue of $3,500,000, in the second year they had a revenue of $9,200,000.

Recovery Year	3-year class	5-year class	7-year class	10-year class
1	33.33	20.00	14.29	10.00
2	44.45	32.00	24.49	18.00
3	*14.81	19.20	17.49	14.40
4	7.41	*11.52	12.49	11.52
5		11.52	*8.93	9.22
6		5.76	8.92	7.37
7			8.93	*6.55
8			4.46	6.55
9				6.56
10				6.55
11				3.28
12-15	In 15-year 12-15 alternate 5.90 & 5.91			
16				
17-20	In 20-year 17-20 alternate 4.461 & 4.462			
21				

IMG 8.3

Kunito had expenses of $125,000 for the 1st year and $350,000 for the second year. Determine the following:

a. The depreciation for year 1 and year 2
b. The taxable income for year 1 and year 2
c. The combined incremental tax rate if the state income tax rate is 7.5% and the federal income tax rate is 21%

8. Solve problem 7 considering that instead of a car fleet the company invested in an asset with a life of (a) 3 years (b) 7 years.

Appendix A: Compound Interest Rate Tables

Interest Rate	0.25%

	single payment		Uniform payment series				Arithmetic Gradient	
n	F/P	P/F	A/F	A/P	F/A	P/A	A/G	P/G
1	1.0025	0.9975	1.0000	1.0025	1.0000	0.9975	0.0000	0.0000
2	1.0050	0.9950	0.4994	0.5019	2.0025	1.9925	0.4994	0.9950
3	1.0075	0.9925	0.3325	0.3350	3.0075	2.9851	0.9983	2.9801
4	1.0100	0.9901	0.2491	0.2516	4.0150	3.9751	1.4969	5.9503
5	1.0126	0.9876	0.1990	0.2015	5.0251	4.9627	1.9950	9.9007
6	1.0151	0.9851	0.1656	0.1681	6.0376	5.9478	2.4927	14.8263
7	1.0176	0.9827	0.1418	0.1443	7.0527	6.9305	2.9900	20.7223
8	1.0202	0.9802	0.1239	0.1264	8.0704	7.9107	3.4869	27.5839
9	1.0227	0.9778	0.1100	0.1125	9.0905	8.8885	3.9834	35.4061
10	1.0253	0.9753	0.0989	0.1014	10.1133	9.8639	4.4794	44.1842
11	1.0278	0.9729	0.0898	0.0923	11.1385	10.8368	4.9750	53.9133
12	1.0304	0.9705	0.0822	0.0847	12.1664	11.8073	5.4702	64.5886
13	1.0330	0.9681	0.0758	0.0783	13.1968	12.7753	5.9650	76.2053
14	1.0356	0.9656	0.0703	0.0728	14.2298	13.7410	6.4594	88.7587
15	1.0382	0.9632	0.0655	0.0680	15.2654	14.7042	6.9534	102.2441
16	1.0408	0.9608	0.0613	0.0638	16.3035	15.6650	7.4469	116.6567
17	1.0434	0.9584	0.0577	0.0602	17.3443	16.6235	7.9401	131.9917
19	1.0486	0.9537	0.0515	0.0540	19.4336	18.5332	8.9251	165.4106
20	1.0512	0.9513	0.0488	0.0513	20.4822	19.4845	9.4170	183.4851
21	1.0538	0.9489	0.0464	0.0489	21.5334	20.4334	9.9085	202.4634
23	1.0591	0.9442	0.0423	0.0448	23.6437	22.3241	10.8901	243.1131
24	1.0618	0.9418	0.0405	0.0430	24.7028	23.2660	11.3804	264.7753
25	1.0644	0.9395	0.0388	0.0413	25.7646	24.2055	11.8702	287.3230
27	1.0697	0.9348	0.0358	0.0383	27.8961	26.0774	12.8485	335.0566
28	1.0724	0.9325	0.0345	0.0370	28.9658	27.0099	13.3371	360.2334
29	1.0751	0.9301	0.0333	0.0358	30.0382	27.9400	13.8252	386.2776
30	1.0778	0.9278	0.0321	0.0346	31.1133	28.8679	14.3130	413.1847
35	1.0913	0.9163	0.0274	0.0299	36.5292	33.4724	16.7454	560.5076
36	1.0941	0.9140	0.0266	0.0291	37.6206	34.3865	17.2306	592.4988
40	1.1050	0.9050	0.0238	0.0263	42.0132	38.0199	19.1673	728.7399
45	1.1189	0.8937	0.0210	0.0235	47.5661	42.5109	21.5789	917.3400
50	1.1330	0.8826	0.0188	0.0213	53.1887	46.9462	23.9802	1125.7767
60	1.1616	0.8609	0.0155	0.0180	64.6467	55.6524	28.7514	1600.0845
65	1.1762	0.8502	0.0142	0.0167	70.4839	59.9246	31.1215	1864.9427
70	1.1910	0.8396	0.0131	0.0156	76.3944	64.1439	33.4812	2147.6111
80	1.2211	0.8189	0.0113	0.0138	88.4392	72.4260	38.1694	2764.4568
90	1.2520	0.7987	0.0099	0.0124	100.7885	80.5038	42.8162	3446.8700
100	1.2836	0.7790	0.0088	0.0113	113.4500	88.3825	47.4216	4191.2417
110	1.3161	0.7598	0.0079	0.0104	126.4316	96.0669	51.9857	4994.1014
120	1.3494	0.7411	0.0072	0.0097	139.7414	103.5618	56.5084	5852.1116

Interest Rate	0.50%

	single payment		Uniform payment series				Arithmetic Gradient	
n	F/P	P/F	A/F	A/P	F/A	P/A	A/G	P/G
1	1.0050	0.9950	1.0000	1.0050	1.0000	0.9950	0.0000	0.0000
2	1.0100	0.9901	0.4988	0.5038	2.0050	1.9851	0.4988	0.9901
3	1.0151	0.9851	0.3317	0.3367	3.0150	2.9702	0.9967	2.9604
4	1.0202	0.9802	0.2481	0.2531	4.0301	3.9505	1.4938	5.9011
5	1.0253	0.9754	0.1980	0.2030	5.0503	4.9259	1.9900	9.8026
6	1.0304	0.9705	0.1646	0.1696	6.0755	5.8964	2.4855	14.6552
7	1.0355	0.9657	0.1407	0.1457	7.1059	6.8621	2.9801	20.4493
8	1.0407	0.9609	0.1228	0.1278	8.1414	7.8230	3.4738	27.1755
9	1.0459	0.9561	0.1089	0.1139	9.1821	8.7791	3.9668	34.8244
10	1.0511	0.9513	0.0978	0.1028	10.2280	9.7304	4.4589	43.3865
11	1.0564	0.9466	0.0887	0.0937	11.2792	10.6770	4.9501	52.8526
12	1.0617	0.9419	0.0811	0.0861	12.3356	11.6189	5.4406	63.2136
13	1.0670	0.9372	0.0746	0.0796	13.3972	12.5562	5.9302	74.4602
14	1.0723	0.9326	0.0691	0.0741	14.4642	13.4887	6.4190	86.5835
15	1.0777	0.9279	0.0644	0.0694	15.5365	14.4166	6.9069	99.5743
16	1.0831	0.9233	0.0602	0.0652	16.6142	15.3399	7.3940	113.4238
17	1.0885	0.9187	0.0565	0.0615	17.6973	16.2586	7.8803	128.1231
19	1.0994	0.9096	0.0503	0.0553	19.8797	18.0824	8.8504	160.0360
20	1.1049	0.9051	0.0477	0.0527	20.9791	18.9874	9.3342	177.2322
21	1.1104	0.9006	0.0453	0.0503	22.0840	19.8880	9.8172	195.2434
23	1.1216	0.8916	0.0411	0.0461	24.3104	21.6757	10.7806	233.6768
24	1.1272	0.8872	0.0393	0.0443	25.4320	22.5629	11.2611	254.0820
25	1.1328	0.8828	0.0377	0.0427	26.5591	23.4456	11.7407	275.2686
27	1.1442	0.8740	0.0347	0.0397	28.8304	25.1980	12.6975	319.9523
28	1.1499	0.8697	0.0334	0.0384	29.9745	26.0677	13.1747	343.4332
29	1.1556	0.8653	0.0321	0.0371	31.1244	26.9330	13.6510	367.6625
30	1.1614	0.8610	0.0310	0.0360	32.2800	27.7941	14.1265	392.6324
35	1.1907	0.8398	0.0262	0.0312	38.1454	32.0354	16.4915	528.3123
36	1.1967	0.8356	0.0254	0.0304	39.3361	32.8710	16.9621	557.5598
40	1.2208	0.8191	0.0226	0.0276	44.1588	36.1722	18.8359	681.3347
45	1.2516	0.7990	0.0199	0.0249	50.3242	40.2072	21.1595	850.7631
50	1.2832	0.7793	0.0177	0.0227	56.6452	44.1428	23.4624	1035.6966
60	1.3489	0.7414	0.0143	0.0193	69.7700	51.7256	28.0064	1448.6458
65	1.3829	0.7231	0.0131	0.0181	76.5821	55.3775	30.2475	1675.0272
70	1.4178	0.7053	0.0120	0.0170	83.5661	58.9394	32.4680	1913.6427
80	1.4903	0.6710	0.0102	0.0152	98.0677	65.8023	36.8474	2424.6455
90	1.5666	0.6383	0.0088	0.0138	113.3109	72.3313	41.1451	2976.0769
100	1.6467	0.6073	0.0077	0.0127	129.3337	78.5426	45.3613	3562.7934
110	1.7309	0.5777	0.0068	0.0118	146.1759	84.4518	49.4964	4180.0565
120	1.8194	0.5496	0.0061	0.0111	163.8793	90.0735	53.5508	4823.5051

Interest Rate	0.75%

	single payment		Uniform payment series				Arithmetic Gradient	
n	F/P	P/F	A/F	A/P	F/A	P/A	A/G	P/G
1	1.0075	0.9926	1.0000	1.0075	1.0000	0.9926	0.0000	0.0000
2	1.0151	0.9852	0.4981	0.5056	2.0075	1.9777	0.4981	0.9852
3	1.0227	0.9778	0.3308	0.3383	3.0226	2.9556	0.9950	2.9408
4	1.0303	0.9706	0.2472	0.2547	4.0452	3.9261	1.4907	5.8525
5	1.0381	0.9633	0.1970	0.2045	5.0756	4.8894	1.9851	9.7058
6	1.0459	0.9562	0.1636	0.1711	6.1136	5.8456	2.4782	14.4866
7	1.0537	0.9490	0.1397	0.1472	7.1595	6.7946	2.9701	20.1808
8	1.0616	0.9420	0.1218	0.1293	8.2132	7.7366	3.4608	26.7747
9	1.0696	0.9350	0.1078	0.1153	9.2748	8.6716	3.9502	34.2544
10	1.0776	0.9280	0.0967	0.1042	10.3443	9.5996	4.4384	42.6064
11	1.0857	0.9211	0.0876	0.0951	11.4219	10.5207	4.9253	51.8174
12	1.0938	0.9142	0.0800	0.0875	12.5076	11.4349	5.4110	61.8740
13	1.1020	0.9074	0.0735	0.0810	13.6014	12.3423	5.8954	72.7632
14	1.1103	0.9007	0.0680	0.0755	14.7034	13.2430	6.3786	84.4720
15	1.1186	0.8940	0.0632	0.0707	15.8137	14.1370	6.8606	96.9876
16	1.1270	0.8873	0.0591	0.0666	16.9323	15.0243	7.3413	110.2973
17	1.1354	0.8807	0.0554	0.0629	18.0593	15.9050	7.8207	124.3887
19	1.1525	0.8676	0.0492	0.0567	20.3387	17.6468	8.7759	154.8671
20	1.1612	0.8612	0.0465	0.0540	21.4912	18.5080	9.2516	171.2297
21	1.1699	0.8548	0.0441	0.0516	22.6524	19.3628	9.7261	188.3253
23	1.1875	0.8421	0.0400	0.0475	25.0010	21.0533	10.6714	224.6682
24	1.1964	0.8358	0.0382	0.0457	26.1885	21.8891	11.1422	243.8923
25	1.2054	0.8296	0.0365	0.0440	27.3849	22.7188	11.6117	263.8029
27	1.2235	0.8173	0.0336	0.0411	29.8047	24.3595	12.5470	305.6387
28	1.2327	0.8112	0.0322	0.0397	31.0282	25.1707	13.0128	327.5416
29	1.2420	0.8052	0.0310	0.0385	32.2609	25.9759	13.4774	350.0867
30	1.2513	0.7992	0.0298	0.0373	33.5029	26.7751	13.9407	373.2631
35	1.2989	0.7699	0.0251	0.0326	39.8538	30.6827	16.2387	498.2471
36	1.3086	0.7641	0.0243	0.0318	41.1527	31.4468	16.6946	524.9924
40	1.3483	0.7416	0.0215	0.0290	46.4465	34.4469	18.5058	637.4693
45	1.3997	0.7145	0.0188	0.0263	53.2901	38.0732	20.7421	789.7173
50	1.4530	0.6883	0.0166	0.0241	60.3943	41.5664	22.9476	953.8486
60	1.5657	0.6387	0.0133	0.0208	75.4241	48.1734	27.2665	1313.5189
65	1.6253	0.6153	0.0120	0.0195	83.3709	51.2963	29.3801	1507.0910
70	1.6872	0.5927	0.0109	0.0184	91.6201	54.3046	31.4634	1708.6065
80	1.8180	0.5500	0.0092	0.0167	109.0725	59.9944	35.5391	2132.1472
90	1.9591	0.5104	0.0078	0.0153	127.8790	65.2746	39.4946	2577.9961
100	2.1111	0.4737	0.0068	0.0143	148.1445	70.1746	43.3311	3040.7453
110	2.2749	0.4396	0.0059	0.0134	169.9823	74.7219	47.0498	3515.6503
120	2.4514	0.4079	0.0052	0.0127	193.5143	78.9417	50.6521	3998.5621

Interest Rate	1.00%

	single payment		Uniform payment series				Arithmetic Gradient	
n	F/P	P/F	A/F	A/P	F/A	P/A	A/G	P/G
1	1.0100	0.9901	1.0000	1.0100	1.0000	0.9901	0.0000	0.0000
2	1.0201	0.9803	0.4975	0.5075	2.0100	1.9704	0.4975	0.9803
3	1.0303	0.9706	0.3300	0.3400	3.0301	2.9410	0.9934	2.9215
4	1.0406	0.9610	0.2463	0.2563	4.0604	3.9020	1.4876	5.8044
5	1.0510	0.9515	0.1960	0.2060	5.1010	4.8534	1.9801	9.6103
6	1.0615	0.9420	0.1625	0.1725	6.1520	5.7955	2.4710	14.3205
7	1.0721	0.9327	0.1386	0.1486	7.2135	6.7282	2.9602	19.9168
8	1.0829	0.9235	0.1207	0.1307	8.2857	7.6517	3.4478	26.3812
9	1.0937	0.9143	0.1067	0.1167	9.3685	8.5660	3.9337	33.6959
10	1.1046	0.9053	0.0956	0.1056	10.4622	9.4713	4.4179	41.8435
11	1.1157	0.8963	0.0865	0.0965	11.5668	10.3676	4.9005	50.8067
12	1.1268	0.8874	0.0788	0.0888	12.6825	11.2551	5.3815	60.5687
13	1.1381	0.8787	0.0724	0.0824	13.8093	12.1337	5.8607	71.1126
14	1.1495	0.8700	0.0669	0.0769	14.9474	13.0037	6.3384	82.4221
15	1.1610	0.8613	0.0621	0.0721	16.0969	13.8651	6.8143	94.4810
16	1.1726	0.8528	0.0579	0.0679	17.2579	14.7179	7.2886	107.2734
17	1.1843	0.8444	0.0543	0.0643	18.4304	15.5623	7.7613	120.7834
19	1.2081	0.8277	0.0481	0.0581	20.8109	17.2260	8.7017	149.8950
20	1.2202	0.8195	0.0454	0.0554	22.0190	18.0456	9.1694	165.4664
21	1.2324	0.8114	0.0430	0.0530	23.2392	18.8570	9.6354	181.6950
23	1.2572	0.7954	0.0389	0.0489	25.7163	20.4558	10.5626	216.0660
24	1.2697	0.7876	0.0371	0.0471	26.9735	21.2434	11.0237	234.1800
25	1.2824	0.7798	0.0354	0.0454	28.2432	22.0232	11.4831	252.8945
27	1.3082	0.7644	0.0324	0.0424	30.8209	23.5596	12.3971	292.0702
28	1.3213	0.7568	0.0311	0.0411	32.1291	24.3164	12.8516	312.5047
29	1.3345	0.7493	0.0299	0.0399	33.4504	25.0658	13.3044	333.4863
30	1.3478	0.7419	0.0287	0.0387	34.7849	25.8077	13.7557	355.0021
35	1.4166	0.7059	0.0240	0.0340	41.6603	29.4086	15.9871	470.1583
36	1.4308	0.6989	0.0232	0.0332	43.0769	30.1075	16.4285	494.6207
40	1.4889	0.6717	0.0205	0.0305	48.8864	32.8347	18.1776	596.8561
45	1.5648	0.6391	0.0177	0.0277	56.4811	36.0945	20.3273	733.7037
50	1.6446	0.6080	0.0155	0.0255	64.4632	39.1961	22.4363	879.4176
60	1.8167	0.5504	0.0122	0.0222	81.6697	44.9550	26.5333	1192.8061
65	1.9094	0.5237	0.0110	0.0210	90.9366	47.6266	28.5217	1358.3903
70	2.0068	0.4983	0.0099	0.0199	100.6763	50.1685	30.4703	1528.6474
80	2.2167	0.4511	0.0082	0.0182	121.6715	54.8882	34.2492	1879.8771
90	2.4486	0.4084	0.0069	0.0169	144.8633	59.1609	37.8724	2240.5675
100	2.7048	0.3697	0.0059	0.0159	170.4814	63.0289	41.3426	2605.7758
110	2.9878	0.3347	0.0050	0.0150	198.7797	66.5305	44.6624	2971.4105
120	3.3004	0.3030	0.0043	0.0143	230.0387	69.7005	47.8349	3334.1148

Interest Rate	1.25%

	single payment		Uniform payment series				Arithmetic Gradient	
n	F/P	P/F	A/F	A/P	F/A	P/A	A/G	P/G
1	1.0125	0.9877	1.0000	1.0125	1.0000	0.9877	0.0000	0.0000
2	1.0252	0.9755	0.4969	0.5094	2.0125	1.9631	0.4969	0.9755
3	1.0380	0.9634	0.3292	0.3417	3.0377	2.9265	0.9917	2.9023
4	1.0509	0.9515	0.2454	0.2579	4.0756	3.8781	1.4845	5.7569
5	1.0641	0.9398	0.1951	0.2076	5.1266	4.8178	1.9752	9.5160
6	1.0774	0.9282	0.1615	0.1740	6.1907	5.7460	2.4638	14.1569
7	1.0909	0.9167	0.1376	0.1501	7.2680	6.6627	2.9503	19.6571
8	1.1045	0.9054	0.1196	0.1321	8.3589	7.5681	3.4348	25.9949
9	1.1183	0.8942	0.1057	0.1182	9.4634	8.4623	3.9172	33.1487
10	1.1323	0.8832	0.0945	0.1070	10.5817	9.3455	4.3975	41.0973
11	1.1464	0.8723	0.0854	0.0979	11.7139	10.2178	4.8758	49.8201
12	1.1608	0.8615	0.0778	0.0903	12.8604	11.0793	5.3520	59.2967
13	1.1753	0.8509	0.0713	0.0838	14.0211	11.9302	5.8262	69.5072
14	1.1900	0.8404	0.0658	0.0783	15.1964	12.7706	6.2982	80.4320
15	1.2048	0.8300	0.0610	0.0735	16.3863	13.6005	6.7682	92.0519
16	1.2199	0.8197	0.0568	0.0693	17.5912	14.4203	7.2362	104.3481
17	1.2351	0.8096	0.0532	0.0657	18.8111	15.2299	7.7021	117.3021
19	1.2662	0.7898	0.0470	0.0595	21.2968	16.8193	8.6277	145.1115
20	1.2820	0.7800	0.0443	0.0568	22.5630	17.5993	9.0874	159.9316
21	1.2981	0.7704	0.0419	0.0544	23.8450	18.3697	9.5450	175.3392
23	1.3307	0.7515	0.0378	0.0503	26.4574	19.8820	10.4542	207.8499
24	1.3474	0.7422	0.0360	0.0485	27.7881	20.6242	10.9056	224.9204
25	1.3642	0.7330	0.0343	0.0468	29.1354	21.3573	11.3551	242.5132
27	1.3985	0.7150	0.0314	0.0439	31.8809	22.7963	12.2478	279.2040
28	1.4160	0.7062	0.0300	0.0425	33.2794	23.5025	12.6911	298.2719
29	1.4337	0.6975	0.0288	0.0413	34.6954	24.2000	13.1323	317.8019
30	1.4516	0.6889	0.0277	0.0402	36.1291	24.8889	13.5715	337.7797
35	1.5446	0.6474	0.0230	0.0355	43.5709	28.2079	15.7369	443.9037
36	1.5639	0.6394	0.0222	0.0347	45.1155	28.8473	16.1639	466.2830
40	1.6436	0.6084	0.0194	0.0319	51.4896	31.3269	17.8515	559.2320
45	1.7489	0.5718	0.0167	0.0292	59.9157	34.2582	19.9156	682.2710
50	1.8610	0.5373	0.0145	0.0270	68.8818	37.0129	21.9295	811.6738
60	2.1072	0.4746	0.0113	0.0238	88.5745	42.0346	25.8083	1084.8429
65	2.2422	0.4460	0.0101	0.0226	99.3771	44.3210	27.6741	1226.5421
70	2.3859	0.4191	0.0090	0.0215	110.8720	46.4697	29.4913	1370.4513
80	2.7015	0.3702	0.0073	0.0198	136.1188	50.3867	32.9822	1661.8651
90	3.0588	0.3269	0.0061	0.0186	164.7050	53.8461	36.2855	1953.8303
100	3.4634	0.2887	0.0051	0.0176	197.0723	56.9013	39.4058	2242.2411
110	3.9215	0.2550	0.0043	0.0168	233.7209	59.5997	42.3483	2523.9437
120	4.4402	0.2252	0.0036	0.0161	275.2171	61.9828	45.1184	2796.5694

Interest Rate	1.50%

	single payment		Uniform payment series				Arithmetic Gradient	
n	F/P	P/F	A/F	A/P	F/A	P/A	A/G	P/G
1	1.0150	0.9852	1.0000	1.0150	1.0000	0.9852	0.0000	0.0000
2	1.0302	0.9707	0.4963	0.5113	2.0150	1.9559	0.4963	0.9707
3	1.0457	0.9563	0.3284	0.3434	3.0452	2.9122	0.9901	2.8833
4	1.0614	0.9422	0.2444	0.2594	4.0909	3.8544	1.4814	5.7098
5	1.0773	0.9283	0.1941	0.2091	5.1523	4.7826	1.9702	9.4229
6	1.0934	0.9145	0.1605	0.1755	6.2296	5.6972	2.4566	13.9956
7	1.1098	0.9010	0.1366	0.1516	7.3230	6.5982	2.9405	19.4018
8	1.1265	0.8877	0.1186	0.1336	8.4328	7.4859	3.4219	25.6157
9	1.1434	0.8746	0.1046	0.1196	9.5593	8.3605	3.9008	32.6125
10	1.1605	0.8617	0.0934	0.1084	10.7027	9.2222	4.3772	40.3675
11	1.1779	0.8489	0.0843	0.0993	11.8633	10.0711	4.8512	48.8568
12	1.1956	0.8364	0.0767	0.0917	13.0412	10.9075	5.3227	58.0571
13	1.2136	0.8240	0.0702	0.0852	14.2368	11.7315	5.7917	67.9454
14	1.2318	0.8118	0.0647	0.0797	15.4504	12.5434	6.2582	78.4994
15	1.2502	0.7999	0.0599	0.0749	16.6821	13.3432	6.7223	89.6974
16	1.2690	0.7880	0.0558	0.0708	17.9324	14.1313	7.1839	101.5178
17	1.2880	0.7764	0.0521	0.0671	19.2014	14.9076	7.6431	113.9400
19	1.3270	0.7536	0.0459	0.0609	21.7967	16.4262	8.5539	140.5084
20	1.3469	0.7425	0.0432	0.0582	23.1237	17.1686	9.0057	154.6154
21	1.3671	0.7315	0.0409	0.0559	24.4705	17.9001	9.4550	169.2453
23	1.4084	0.7100	0.0367	0.0517	27.2251	19.3309	10.3462	200.0006
24	1.4295	0.6995	0.0349	0.0499	28.6335	20.0304	10.7881	216.0901
25	1.4509	0.6892	0.0333	0.0483	30.0630	20.7196	11.2276	232.6310
27	1.4948	0.6690	0.0303	0.0453	32.9867	22.0676	12.0992	267.0002
28	1.5172	0.6591	0.0290	0.0440	34.4815	22.7267	12.5313	284.7958
29	1.5400	0.6494	0.0278	0.0428	35.9987	23.3761	12.9610	302.9779
30	1.5631	0.6398	0.0266	0.0416	37.5387	24.0158	13.3883	321.5310
35	1.6839	0.5939	0.0219	0.0369	45.5921	27.0756	15.4882	419.3521
36	1.7091	0.5851	0.0212	0.0362	47.2760	27.6607	15.9009	439.8303
40	1.8140	0.5513	0.0184	0.0334	54.2679	29.9158	17.5277	524.3568
45	1.9542	0.5117	0.0157	0.0307	63.6142	32.5523	19.5074	635.0110
50	2.1052	0.4750	0.0136	0.0286	73.6828	34.9997	21.4277	749.9636
60	2.4432	0.4093	0.0104	0.0254	96.2147	39.3803	25.0930	988.1674
65	2.6320	0.3799	0.0092	0.0242	108.8028	41.3378	26.8393	1109.4752
70	2.8355	0.3527	0.0082	0.0232	122.3638	43.1549	28.5290	1231.1658
80	3.2907	0.3039	0.0065	0.0215	152.7109	46.4073	31.7423	1473.0741
90	3.8189	0.2619	0.0053	0.0203	187.9299	49.2099	34.7399	1709.5439
100	4.4320	0.2256	0.0044	0.0194	228.8030	51.6247	37.5295	1937.4506
110	5.1436	0.1944	0.0036	0.0186	276.2380	53.7055	40.1195	2154.6383
120	5.9693	0.1675	0.0030	0.0180	331.2882	55.4985	42.5185	2359.7114

Interest Rate	1.75%

	single payment		Uniform payment series				Arithmetic Gradient	
n	F/P	P/F	A/F	A/P	F/A	P/A	A/G	P/G
1	1.0175	0.9828	1.0000	1.0175	1.0000	0.9828	0.0000	0.0000
2	1.0353	0.9659	0.4957	0.5132	2.0175	1.9487	0.4957	0.9659
3	1.0534	0.9493	0.3276	0.3451	3.0528	2.8980	0.9884	2.8645
4	1.0719	0.9330	0.2435	0.2610	4.1062	3.8309	1.4783	5.6633
5	1.0906	0.9169	0.1931	0.2106	5.1781	4.7479	1.9653	9.3310
6	1.1097	0.9011	0.1595	0.1770	6.2687	5.6490	2.4494	13.8367
7	1.1291	0.8856	0.1355	0.1530	7.3784	6.5346	2.9306	19.1506
8	1.1489	0.8704	0.1175	0.1350	8.5075	7.4051	3.4089	25.2435
9	1.1690	0.8554	0.1036	0.1211	9.6564	8.2605	3.8844	32.0870
10	1.1894	0.8407	0.0924	0.1099	10.8254	9.1012	4.3569	39.6535
11	1.2103	0.8263	0.0832	0.1007	12.0148	9.9275	4.8266	47.9162
12	1.2314	0.8121	0.0756	0.0931	13.2251	10.7395	5.2934	56.8489
13	1.2530	0.7981	0.0692	0.0867	14.4565	11.5376	5.7573	66.4260
14	1.2749	0.7844	0.0637	0.0812	15.7095	12.3220	6.2184	76.6227
15	1.2972	0.7709	0.0589	0.0764	16.9844	13.0929	6.6765	87.4149
16	1.3199	0.7576	0.0547	0.0722	18.2817	13.8505	7.1318	98.7792
17	1.3430	0.7446	0.0510	0.0685	19.6016	14.5951	7.5842	110.6926
19	1.3904	0.7192	0.0448	0.0623	22.3112	16.0461	8.4805	136.0783
20	1.4148	0.7068	0.0422	0.0597	23.7016	16.7529	8.9243	149.5080
21	1.4395	0.6947	0.0398	0.0573	25.1164	17.4475	9.3653	163.4013
23	1.4904	0.6710	0.0357	0.0532	28.0207	18.8012	10.2387	192.5000
24	1.5164	0.6594	0.0339	0.0514	29.5110	19.4607	10.6711	207.6671
25	1.5430	0.6481	0.0322	0.0497	31.0275	20.1088	11.1007	223.2214
27	1.5975	0.6260	0.0293	0.0468	34.1404	21.3717	11.9513	255.4210
28	1.6254	0.6152	0.0280	0.0455	35.7379	21.9870	12.3724	272.0321
29	1.6539	0.6046	0.0268	0.0443	37.3633	22.5916	12.7907	288.9623
30	1.6828	0.5942	0.0256	0.0431	39.0172	23.1858	13.2061	306.1954
35	1.8353	0.5449	0.0210	0.0385	47.7308	26.0073	15.2412	396.3824
36	1.8674	0.5355	0.0202	0.0377	49.5661	26.5428	15.6399	415.1250
40	2.0016	0.4996	0.0175	0.0350	57.2341	28.5942	17.2066	492.0109
45	2.1830	0.4581	0.0148	0.0323	67.5986	30.9663	19.1032	591.5540
50	2.3808	0.4200	0.0127	0.0302	78.9022	33.1412	20.9317	693.7010
60	2.8318	0.3531	0.0096	0.0271	104.6752	36.9640	24.3885	901.4954
65	3.0884	0.3238	0.0084	0.0259	119.3386	38.6406	26.0189	1005.3872
70	3.3683	0.2969	0.0074	0.0249	135.3308	40.1779	27.5856	1108.3333
80	4.0064	0.2496	0.0058	0.0233	171.7938	42.8799	30.5329	1309.2482
90	4.7654	0.2098	0.0046	0.0221	215.1646	45.1516	33.2409	1500.8798
100	5.6682	0.1764	0.0037	0.0212	266.7518	47.0615	35.7211	1681.0886
110	6.7420	0.1483	0.0030	0.0205	328.1118	48.6671	37.9856	1848.6521
120	8.0192	0.1247	0.0025	0.0200	401.0962	50.0171	40.0469	2003.0269

Interest Rate	2.00%

	single payment		Uniform payment series				Arithmetic Gradient	
n	F/P	P/F	A/F	A/P	F/A	P/A	A/G	P/G
1	1.0200	0.9804	1.0000	1.0200	1.0000	0.9804	0.0000	0.0000
2	1.0404	0.9612	0.4950	0.5150	2.0200	1.9416	0.4950	0.9612
3	1.0612	0.9423	0.3268	0.3468	3.0604	2.8839	0.9868	2.8458
4	1.0824	0.9238	0.2426	0.2626	4.1216	3.8077	1.4752	5.6173
5	1.1041	0.9057	0.1922	0.2122	5.2040	4.7135	1.9604	9.2403
6	1.1262	0.8880	0.1585	0.1785	6.3081	5.6014	2.4423	13.6801
7	1.1487	0.8706	0.1345	0.1545	7.4343	6.4720	2.9208	18.9035
8	1.1717	0.8535	0.1165	0.1365	8.5830	7.3255	3.3961	24.8779
9	1.1951	0.8368	0.1025	0.1225	9.7546	8.1622	3.8681	31.5720
10	1.2190	0.8203	0.0913	0.1113	10.9497	8.9826	4.3367	38.9551
11	1.2434	0.8043	0.0822	0.1022	12.1687	9.7868	4.8021	46.9977
12	1.2682	0.7885	0.0746	0.0946	13.4121	10.5753	5.2642	55.6712
13	1.2936	0.7730	0.0681	0.0881	14.6803	11.3484	5.7231	64.9475
14	1.3195	0.7579	0.0626	0.0826	15.9739	12.1062	6.1786	74.7999
15	1.3459	0.7430	0.0578	0.0778	17.2934	12.8493	6.6309	85.2021
16	1.3728	0.7284	0.0537	0.0737	18.6393	13.5777	7.0799	96.1288
17	1.4002	0.7142	0.0500	0.0700	20.0121	14.2919	7.5256	107.5554
19	1.4568	0.6864	0.0438	0.0638	22.8406	15.6785	8.4073	131.8139
20	1.4859	0.6730	0.0412	0.0612	24.2974	16.3514	8.8433	144.6003
21	1.5157	0.6598	0.0388	0.0588	25.7833	17.0112	9.2760	157.7959
23	1.5769	0.6342	0.0347	0.0547	28.8450	18.2922	10.1317	185.3309
24	1.6084	0.6217	0.0329	0.0529	30.4219	18.9139	10.5547	199.6305
25	1.6406	0.6095	0.0312	0.0512	32.0303	19.5235	10.9745	214.2592
27	1.7069	0.5859	0.0283	0.0483	35.3443	20.7069	11.8043	244.4311
28	1.7410	0.5744	0.0270	0.0470	37.0512	21.2813	12.2145	259.9392
29	1.7758	0.5631	0.0258	0.0458	38.7922	21.8444	12.6214	275.7064
30	1.8114	0.5521	0.0246	0.0446	40.5681	22.3965	13.0251	291.7164
35	1.9999	0.5000	0.0200	0.0400	49.9945	24.9986	14.9961	374.8826
36	2.0399	0.4902	0.0192	0.0392	51.9944	25.4888	15.3809	392.0405
40	2.2080	0.4529	0.0166	0.0366	60.4020	27.3555	16.8885	461.9931
45	2.4379	0.4102	0.0139	0.0339	71.8927	29.4902	18.7034	551.5652
50	2.6916	0.3715	0.0118	0.0318	84.5794	31.4236	20.4420	642.3606
60	3.2810	0.3048	0.0088	0.0288	114.0515	34.7609	23.6961	823.6975
65	3.6225	0.2761	0.0076	0.0276	131.1262	36.1975	25.2147	912.7085
70	3.9996	0.2500	0.0067	0.0267	149.9779	37.4986	26.6632	999.8343
80	4.8754	0.2051	0.0052	0.0252	193.7720	39.7445	29.3572	1166.7868
90	5.9431	0.1683	0.0040	0.0240	247.1567	41.5869	31.7929	1322.1701
100	7.2446	0.1380	0.0032	0.0232	312.2323	43.0984	33.9863	1464.7527
110	8.8312	0.1132	0.0026	0.0226	391.5592	44.3382	35.9536	1594.1191
120	10.7652	0.0929	0.0020	0.0220	488.2582	45.3554	37.7114	1710.4160

Interest Rate	2.50%

	single payment		Uniform payment series				Arithmetic Gradient	
n	F/P	P/F	A/F	A/P	F/A	P/A	A/G	P/G
1	1.0250	0.9756	1.0000	1.0250	1.0000	0.9756	0.0000	0.0000
2	1.0506	0.9518	0.4938	0.5188	2.0250	1.9274	0.4938	0.9518
3	1.0769	0.9286	0.3251	0.3501	3.0756	2.8560	0.9835	2.8090
4	1.1038	0.9060	0.2408	0.2658	4.1525	3.7620	1.4691	5.5269
5	1.1314	0.8839	0.1902	0.2152	5.2563	4.6458	1.9506	9.0623
6	1.1597	0.8623	0.1565	0.1815	6.3877	5.5081	2.4280	13.3738
7	1.1887	0.8413	0.1325	0.1575	7.5474	6.3494	2.9013	18.4214
8	1.2184	0.8207	0.1145	0.1395	8.7361	7.1701	3.3704	24.1666
9	1.2489	0.8007	0.1005	0.1255	9.9545	7.9709	3.8355	30.5724
10	1.2801	0.7812	0.0893	0.1143	11.2034	8.7521	4.2965	37.6032
11	1.3121	0.7621	0.0801	0.1051	12.4835	9.5142	4.7534	45.2246
12	1.3449	0.7436	0.0725	0.0975	13.7956	10.2578	5.2062	53.4038
13	1.3785	0.7254	0.0660	0.0910	15.1404	10.9832	5.6549	62.1088
14	1.4130	0.7077	0.0605	0.0855	16.5190	11.6909	6.0995	71.3093
15	1.4483	0.6905	0.0558	0.0808	17.9319	12.3814	6.5401	80.9758
16	1.4845	0.6736	0.0516	0.0766	19.3802	13.0550	6.9766	91.0801
17	1.5216	0.6572	0.0479	0.0729	20.8647	13.7122	7.4091	101.5953
19	1.5987	0.6255	0.0418	0.0668	23.9460	14.9789	8.2619	123.7546
20	1.6386	0.6103	0.0391	0.0641	25.5447	15.5892	8.6823	135.3497
21	1.6796	0.5954	0.0368	0.0618	27.1833	16.1845	9.0986	147.2575
23	1.7646	0.5667	0.0327	0.0577	30.5844	17.3321	9.9193	171.9230
24	1.8087	0.5529	0.0309	0.0559	32.3490	17.8850	10.3237	184.6391
25	1.8539	0.5394	0.0293	0.0543	34.1578	18.4244	10.7241	197.5845
27	1.9478	0.5134	0.0264	0.0514	37.9120	19.4640	11.5130	224.0887
28	1.9965	0.5009	0.0251	0.0501	39.8598	19.9649	11.9015	237.6124
29	2.0464	0.4887	0.0239	0.0489	41.8563	20.4535	12.2861	251.2949
30	2.0976	0.4767	0.0228	0.0478	43.9027	20.9303	12.6668	265.1205
35	2.3732	0.4214	0.0182	0.0432	54.9282	23.1452	14.5122	335.8868
36	2.4325	0.4111	0.0175	0.0425	57.3014	23.5563	14.8697	350.2751
40	2.6851	0.3724	0.0148	0.0398	67.4026	25.1028	16.2620	408.2220
45	3.0379	0.3292	0.0123	0.0373	81.5161	26.8330	17.9185	480.8070
50	3.4371	0.2909	0.0103	0.0353	97.4843	28.3623	19.4839	552.6081
60	4.3998	0.2273	0.0074	0.0324	135.9916	30.9087	22.3518	690.8656
65	4.9780	0.2009	0.0063	0.0313	159.1183	31.9646	23.6600	756.2806
70	5.6321	0.1776	0.0054	0.0304	185.2841	32.8979	24.8881	818.7643
80	7.2096	0.1387	0.0040	0.0290	248.3827	34.4518	27.1167	934.2181
90	9.2289	0.1084	0.0030	0.0280	329.1543	35.6658	29.0629	1036.5499
100	11.8137	0.0846	0.0023	0.0273	432.5487	36.6141	30.7525	1125.9747
110	15.1226	0.0661	0.0018	0.0268	564.9022	37.3549	32.2110	1203.2417
120	19.3581	0.0517	0.0014	0.0264	734.3260	37.9337	33.4634	1269.3899

Interest Rate	3.00%

	single payment		Uniform payment series				Arithmetic Gradient	
n	F/P	P/F	A/F	A/P	F/A	P/A	A/G	P/G
1	1.0300	0.9709	1.0000	1.0300	1.0000	0.9709	0.0000	0.0000
2	1.0609	0.9426	0.4926	0.5226	2.0300	1.9135	0.4926	0.9426
3	1.0927	0.9151	0.3235	0.3535	3.0909	2.8286	0.9803	2.7729
4	1.1255	0.8885	0.2390	0.2690	4.1836	3.7171	1.4631	5.4383
5	1.1593	0.8626	0.1884	0.2184	5.3091	4.5797	1.9409	8.8888
6	1.1941	0.8375	0.1546	0.1846	6.4684	5.4172	2.4138	13.0762
7	1.2299	0.8131	0.1305	0.1605	7.6625	6.2303	2.8819	17.9547
8	1.2668	0.7894	0.1125	0.1425	8.8923	7.0197	3.3450	23.4806
9	1.3048	0.7664	0.0984	0.1284	10.1591	7.7861	3.8032	29.6119
10	1.3439	0.7441	0.0872	0.1172	11.4639	8.5302	4.2565	36.3088
11	1.3842	0.7224	0.0781	0.1081	12.8078	9.2526	4.7049	43.5330
12	1.4258	0.7014	0.0705	0.1005	14.1920	9.9540	5.1485	51.2482
13	1.4685	0.6810	0.0640	0.0940	15.6178	10.6350	5.5872	59.4196
14	1.5126	0.6611	0.0585	0.0885	17.0863	11.2961	6.0210	68.0141
15	1.5580	0.6419	0.0538	0.0838	18.5989	11.9379	6.4500	77.0002
16	1.6047	0.6232	0.0496	0.0796	20.1569	12.5611	6.8742	86.3477
17	1.6528	0.6050	0.0460	0.0760	21.7616	13.1661	7.2936	96.0280
19	1.7535	0.5703	0.0398	0.0698	25.1169	14.3238	8.1179	116.2788
20	1.8061	0.5537	0.0372	0.0672	26.8704	14.8775	8.5229	126.7987
21	1.8603	0.5375	0.0349	0.0649	28.6765	15.4150	8.9231	137.5496
23	1.9736	0.5067	0.0308	0.0608	32.4529	16.4436	9.7093	159.6566
24	2.0328	0.4919	0.0290	0.0590	34.4265	16.9355	10.0954	170.9711
25	2.0938	0.4776	0.0274	0.0574	36.4593	17.4131	10.4768	182.4336
27	2.2213	0.4502	0.0246	0.0546	40.7096	18.3270	11.2255	205.7309
28	2.2879	0.4371	0.0233	0.0533	42.9309	18.7641	11.5930	217.5320
29	2.3566	0.4243	0.0221	0.0521	45.2189	19.1885	11.9558	229.4137
30	2.4273	0.4120	0.0210	0.0510	47.5754	19.6004	12.3141	241.3613
35	2.8139	0.3554	0.0165	0.0465	60.4621	21.4872	14.0375	301.6267
36	2.8983	0.3450	0.0158	0.0458	63.2759	21.8323	14.3688	313.7028
40	3.2620	0.3066	0.0133	0.0433	75.4013	23.1148	15.6502	361.7499
45	3.7816	0.2644	0.0108	0.0408	92.7199	24.5187	17.1556	420.6325
50	4.3839	0.2281	0.0089	0.0389	112.7969	25.7298	18.5575	477.4803
60	5.8916	0.1697	0.0061	0.0361	163.0534	27.6756	21.0674	583.0526
65	6.8300	0.1464	0.0051	0.0351	194.3328	28.4529	22.1841	631.2010
70	7.9178	0.1263	0.0043	0.0343	230.5941	29.1234	23.2145	676.0869
80	10.6409	0.0940	0.0031	0.0331	321.3630	30.2008	25.0353	756.0865
90	14.3005	0.0699	0.0023	0.0323	443.3489	31.0024	26.5667	823.6302
100	19.2186	0.0520	0.0016	0.0316	607.2877	31.5989	27.8444	879.8540
110	25.8282	0.0387	0.0012	0.0312	827.6078	32.0428	28.9029	926.1284
120	34.7110	0.0288	0.0009	0.0309	1123.6996	32.3730	29.7737	963.8635

Interest Rate	4.00%

	single payment		Uniform payment series				Arithmetic Gradient	
n	F/P	P/F	A/F	A/P	F/A	P/A	A/G	P/G
1	1.0400	0.9615	1.0000	1.0400	1.0000	0.9615	0.0000	0.0000
2	1.0816	0.9246	0.4902	0.5302	2.0400	1.8861	0.4902	0.9246
3	1.1249	0.8890	0.3203	0.3603	3.1216	2.7751	0.9739	2.7025
4	1.1699	0.8548	0.2355	0.2755	4.2465	3.6299	1.4510	5.2670
5	1.2167	0.8219	0.1846	0.2246	5.4163	4.4518	1.9216	8.5547
6	1.2653	0.7903	0.1508	0.1908	6.6330	5.2421	2.3857	12.5062
7	1.3159	0.7599	0.1266	0.1666	7.8983	6.0021	2.8433	17.0657
8	1.3686	0.7307	0.1085	0.1485	9.2142	6.7327	3.2944	22.1806
9	1.4233	0.7026	0.0945	0.1345	10.5828	7.4353	3.7391	27.8013
10	1.4802	0.6756	0.0833	0.1233	12.0061	8.1109	4.1773	33.8814
11	1.5395	0.6496	0.0741	0.1141	13.4864	8.7605	4.6090	40.3772
12	1.6010	0.6246	0.0666	0.1066	15.0258	9.3851	5.0343	47.2477
13	1.6651	0.6006	0.0601	0.1001	16.6268	9.9856	5.4533	54.4546
14	1.7317	0.5775	0.0547	0.0947	18.2919	10.5631	5.8659	61.9618
15	1.8009	0.5553	0.0499	0.0899	20.0236	11.1184	6.2721	69.7355
16	1.8730	0.5339	0.0458	0.0858	21.8245	11.6523	6.6720	77.7441
17	1.9479	0.5134	0.0422	0.0822	23.6975	12.1657	7.0656	85.9581
19	2.1068	0.4746	0.0361	0.0761	27.6712	13.1339	7.8342	102.8933
20	2.1911	0.4564	0.0336	0.0736	29.7781	13.5903	8.2091	111.5647
21	2.2788	0.4388	0.0313	0.0713	31.9692	14.0292	8.5779	120.3414
23	2.4647	0.4057	0.0273	0.0673	36.6179	14.8568	9.2973	138.1284
24	2.5633	0.3901	0.0256	0.0656	39.0826	15.2470	9.6479	147.1012
25	2.6658	0.3751	0.0240	0.0640	41.6459	15.6221	9.9925	156.1040
27	2.8834	0.3468	0.0212	0.0612	47.0842	16.3296	10.6640	174.1385
28	2.9987	0.3335	0.0200	0.0600	49.9676	16.6631	10.9909	183.1424
29	3.1187	0.3207	0.0189	0.0589	52.9663	16.9837	11.3120	192.1206
30	3.2434	0.3083	0.0178	0.0578	56.0849	17.2920	11.6274	201.0618
35	3.9461	0.2534	0.0136	0.0536	73.6522	18.6646	13.1198	244.8768
36	4.1039	0.2437	0.0129	0.0529	77.5983	18.9083	13.4018	253.4052
40	4.8010	0.2083	0.0105	0.0505	95.0255	19.7928	14.4765	286.5303
45	5.8412	0.1712	0.0083	0.0483	121.0294	20.7200	15.7047	325.4028
50	7.1067	0.1407	0.0066	0.0466	152.6671	21.4822	16.8122	361.1638
60	10.5196	0.0951	0.0042	0.0442	237.9907	22.6235	18.6972	422.9966
65	12.7987	0.0781	0.0034	0.0434	294.9684	23.0467	19.4909	449.2014
70	15.5716	0.0642	0.0027	0.0427	364.2905	23.3945	20.1961	472.4789
80	23.0498	0.0434	0.0018	0.0418	551.2450	23.9154	21.3718	511.1161
90	34.1193	0.0293	0.0012	0.0412	827.9833	24.2673	22.2826	540.7369
100	50.5049	0.0198	0.0008	0.0408	1237.6237	24.5050	22.9800	563.1249
110	74.7597	0.0134	0.0005	0.0405	1843.9915	24.6656	23.5087	579.8553
120	110.6626	0.0090	0.0004	0.0404	2741.5640	24.7741	23.9057	592.2428

Interest Rate	5.00%

	single payment		Uniform payment series				Arithmetic Gradient	
n	F/P	P/F	A/F	A/P	F/A	P/A	A/G	P/G
1	1.0500	0.9524	1.0000	1.0500	1.0000	0.9524	0.0000	0.0000
2	1.1025	0.9070	0.4878	0.5378	2.0500	1.8594	0.4878	0.9070
3	1.1576	0.8638	0.3172	0.3672	3.1525	2.7232	0.9675	2.6347
4	1.2155	0.8227	0.2320	0.2820	4.3101	3.5460	1.4391	5.1028
5	1.2763	0.7835	0.1810	0.2310	5.5256	4.3295	1.9025	8.2369
6	1.3401	0.7462	0.1470	0.1970	6.8019	5.0757	2.3579	11.9680
7	1.4071	0.7107	0.1228	0.1728	8.1420	5.7864	2.8052	16.2321
8	1.4775	0.6768	0.1047	0.1547	9.5491	6.4632	3.2445	20.9700
9	1.5513	0.6446	0.0907	0.1407	11.0266	7.1078	3.6758	26.1268
10	1.6289	0.6139	0.0795	0.1295	12.5779	7.7217	4.0991	31.6520
11	1.7103	0.5847	0.0704	0.1204	14.2068	8.3064	4.5144	37.4988
12	1.7959	0.5568	0.0628	0.1128	15.9171	8.8633	4.9219	43.6241
13	1.8856	0.5303	0.0565	0.1065	17.7130	9.3936	5.3215	49.9879
14	1.9799	0.5051	0.0510	0.1010	19.5986	9.8986	5.7133	56.5538
15	2.0789	0.4810	0.0463	0.0963	21.5786	10.3797	6.0973	63.2880
16	2.1829	0.4581	0.0423	0.0923	23.6575	10.8378	6.4736	70.1597
17	2.2920	0.4363	0.0387	0.0887	25.8404	11.2741	6.8423	77.1405
19	2.5270	0.3957	0.0327	0.0827	30.5390	12.0853	7.5569	91.3275
20	2.6533	0.3769	0.0302	0.0802	33.0660	12.4622	7.9030	98.4884
21	2.7860	0.3589	0.0280	0.0780	35.7193	12.8212	8.2416	105.6673
23	3.0715	0.3256	0.0241	0.0741	41.4305	13.4886	8.8971	120.0087
24	3.2251	0.3101	0.0225	0.0725	44.5020	13.7986	9.2140	127.1402
25	3.3864	0.2953	0.0210	0.0710	47.7271	14.0939	9.5238	134.2275
27	3.7335	0.2678	0.0183	0.0683	54.6691	14.6430	10.1224	148.2226
28	3.9201	0.2551	0.0171	0.0671	58.4026	14.8981	10.4114	155.1101
29	4.1161	0.2429	0.0160	0.0660	62.3227	15.1411	10.6936	161.9126
30	4.3219	0.2314	0.0151	0.0651	66.4388	15.3725	10.9691	168.6226
35	5.5160	0.1813	0.0111	0.0611	90.3203	16.3742	12.2498	200.5807
36	5.7918	0.1727	0.0104	0.0604	95.8363	16.5469	12.4872	206.6237
40	7.0400	0.1420	0.0083	0.0583	120.7998	17.1591	13.3775	229.5452
45	8.9850	0.1113	0.0063	0.0563	159.7002	17.7741	14.3644	255.3145
50	11.4674	0.0872	0.0048	0.0548	209.3480	18.2559	15.2233	277.9148
60	18.6792	0.0535	0.0028	0.0528	353.5837	18.9293	16.6062	314.3432
65	23.8399	0.0419	0.0022	0.0522	456.7980	19.1611	17.1541	328.6910
70	30.4264	0.0329	0.0017	0.0517	588.5285	19.3427	17.6212	340.8409
80	49.5614	0.0202	0.0010	0.0510	971.2288	19.5965	18.3526	359.6460
90	80.7304	0.0124	0.0006	0.0506	1594.6073	19.7523	18.8712	372.7488
100	131.5013	0.0076	0.0004	0.0504	2610.0252	19.8479	19.2337	381.7492
110	214.2017	0.0047	0.0002	0.0502	4264.0338	19.9066	19.4841	387.8619
120	348.9120	0.0029	0.0001	0.0501	6958.2397	19.9427	19.6551	391.9751

Interest Rate	6.00%

	single payment		Uniform payment series				Arithmetic Gradient	
n	F/P	P/F	A/F	A/P	F/A	P/A	A/G	P/G
1	1.0600	0.9434	1.0000	1.0600	1.0000	0.9434	0.0000	0.0000
2	1.1236	0.8900	0.4854	0.5454	2.0600	1.8334	0.4854	0.8900
3	1.1910	0.8396	0.3141	0.3741	3.1836	2.6730	0.9612	2.5692
4	1.2625	0.7921	0.2286	0.2886	4.3746	3.4651	1.4272	4.9455
5	1.3382	0.7473	0.1774	0.2374	5.6371	4.2124	1.8836	7.9345
6	1.4185	0.7050	0.1434	0.2034	6.9753	4.9173	2.3304	11.4594
7	1.5036	0.6651	0.1191	0.1791	8.3938	5.5824	2.7676	15.4497
8	1.5938	0.6274	0.1010	0.1610	9.8975	6.2098	3.1952	19.8416
9	1.6895	0.5919	0.0870	0.1470	11.4913	6.8017	3.6133	24.5768
10	1.7908	0.5584	0.0759	0.1359	13.1808	7.3601	4.0220	29.6023
11	1.8983	0.5268	0.0668	0.1268	14.9716	7.8869	4.4213	34.8702
12	2.0122	0.4970	0.0593	0.1193	16.8699	8.3838	4.8113	40.3369
13	2.1329	0.4688	0.0530	0.1130	18.8821	8.8527	5.1920	45.9629
14	2.2609	0.4423	0.0476	0.1076	21.0151	9.2950	5.5635	51.7128
15	2.3966	0.4173	0.0430	0.1030	23.2760	9.7122	5.9260	57.5546
16	2.5404	0.3936	0.0390	0.0990	25.6725	10.1059	6.2794	63.4592
17	2.6928	0.3714	0.0354	0.0954	28.2129	10.4773	6.6240	69.4011
19	3.0256	0.3305	0.0296	0.0896	33.7600	11.1581	7.2867	81.3062
20	3.2071	0.3118	0.0272	0.0872	36.7856	11.4699	7.6051	87.2304
21	3.3996	0.2942	0.0250	0.0850	39.9927	11.7641	7.9151	93.1136
23	3.8197	0.2618	0.0213	0.0813	46.9958	12.3034	8.5099	104.7007
24	4.0489	0.2470	0.0197	0.0797	50.8156	12.5504	8.7951	110.3812
25	4.2919	0.2330	0.0182	0.0782	54.8645	12.7834	9.0722	115.9732
27	4.8223	0.2074	0.0157	0.0757	63.7058	13.2105	9.6029	126.8600
28	5.1117	0.1956	0.0146	0.0746	68.5281	13.4062	9.8568	132.1420
29	5.4184	0.1846	0.0136	0.0736	73.6398	13.5907	10.1032	137.3096
30	5.7435	0.1741	0.0126	0.0726	79.0582	13.7648	10.3422	142.3588
35	7.6861	0.1301	0.0090	0.0690	111.4348	14.4982	11.4319	165.7427
36	8.1473	0.1227	0.0084	0.0684	119.1209	14.6210	11.6298	170.0387
40	10.2857	0.0972	0.0065	0.0665	154.7620	15.0463	12.3590	185.9568
45	13.7646	0.0727	0.0047	0.0647	212.7435	15.4558	13.1413	203.1096
50	18.4202	0.0543	0.0034	0.0634	290.3359	15.7619	13.7964	217.4574
60	32.9877	0.0303	0.0019	0.0619	533.1282	16.1614	14.7909	239.0428
65	44.1450	0.0227	0.0014	0.0614	719.0829	16.2891	15.1601	246.9450
70	59.0759	0.0169	0.0010	0.0610	967.9322	16.3845	15.4613	253.3271
80	105.7960	0.0095	0.0006	0.0606	1746.5999	16.5091	15.9033	262.5493
90	189.4645	0.0053	0.0003	0.0603	3141.0752	16.5787	16.1891	268.3946
100	339.3021	0.0029	0.0002	0.0602	5638.3681	16.6175	16.3711	272.0471
110	607.6384	0.0016	0.0001	0.0601	10110.6392	16.6392	16.4853	274.3035
120	1088.1877	0.0009	0.0001	0.0601	18119.7958	16.6514	16.5563	275.6846

Interest Rate	7.00%

	single payment		Uniform payment series				Arithmetic Gradient	
n	F/P	P/F	A/F	A/P	F/A	P/A	A/G	P/G
1	1.0700	0.9346	1.0000	1.0700	1.0000	0.9346	0.0000	0.0000
2	1.1449	0.8734	0.4831	0.5531	2.0700	1.8080	0.4831	0.8734
3	1.2250	0.8163	0.3111	0.3811	3.2149	2.6243	0.9549	2.5060
4	1.3108	0.7629	0.2252	0.2952	4.4399	3.3872	1.4155	4.7947
5	1.4026	0.7130	0.1739	0.2439	5.7507	4.1002	1.8650	7.6467
6	1.5007	0.6663	0.1398	0.2098	7.1533	4.7665	2.3032	10.9784
7	1.6058	0.6227	0.1156	0.1856	8.6540	5.3893	2.7304	14.7149
8	1.7182	0.5820	0.0975	0.1675	10.2598	5.9713	3.1465	18.7889
9	1.8385	0.5439	0.0835	0.1535	11.9780	6.5152	3.5517	23.1404
10	1.9672	0.5083	0.0724	0.1424	13.8164	7.0236	3.9461	27.7156
11	2.1049	0.4751	0.0634	0.1334	15.7836	7.4987	4.3296	32.4665
12	2.2522	0.4440	0.0559	0.1259	17.8885	7.9427	4.7025	37.3506
13	2.4098	0.4150	0.0497	0.1197	20.1406	8.3577	5.0648	42.3302
14	2.5785	0.3878	0.0443	0.1143	22.5505	8.7455	5.4167	47.3718
15	2.7590	0.3624	0.0398	0.1098	25.1290	9.1079	5.7583	52.4461
16	2.9522	0.3387	0.0359	0.1059	27.8881	9.4466	6.0897	57.5271
17	3.1588	0.3166	0.0324	0.1024	30.8402	9.7632	6.4110	62.5923
19	3.6165	0.2765	0.0268	0.0968	37.3790	10.3356	7.0242	72.5991
20	3.8697	0.2584	0.0244	0.0944	40.9955	10.5940	7.3163	77.5091
21	4.1406	0.2415	0.0223	0.0923	44.8652	10.8355	7.5990	82.3393
23	4.7405	0.2109	0.0187	0.0887	53.4361	11.2722	8.1369	91.7201
24	5.0724	0.1971	0.0172	0.0872	58.1767	11.4693	8.3923	96.2545
25	5.4274	0.1842	0.0158	0.0858	63.2490	11.6536	8.6391	100.6765
27	6.2139	0.1609	0.0134	0.0834	74.4838	11.9867	9.1072	109.1656
28	6.6488	0.1504	0.0124	0.0824	80.6977	12.1371	9.3289	113.2264
29	7.1143	0.1406	0.0114	0.0814	87.3465	12.2777	9.5427	117.1622
30	7.6123	0.1314	0.0106	0.0806	94.4608	12.4090	9.7487	120.9718
35	10.6766	0.0937	0.0072	0.0772	138.2369	12.9477	10.6687	138.1353
36	11.4239	0.0875	0.0067	0.0767	148.9135	13.0352	10.8321	141.1990
40	14.9745	0.0668	0.0050	0.0750	199.6351	13.3317	11.4233	152.2928
45	21.0025	0.0476	0.0035	0.0735	285.7493	13.6055	12.0360	163.7559
50	29.4570	0.0339	0.0025	0.0725	406.5289	13.8007	12.5287	172.9051
60	57.9464	0.0173	0.0012	0.0712	813.5204	14.0392	13.2321	185.7677
65	81.2729	0.0123	0.0009	0.0709	1146.7552	14.1099	13.4760	190.1452
70	113.9894	0.0088	0.0006	0.0706	1614.1342	14.1604	13.6662	193.5185
80	224.2344	0.0045	0.0003	0.0703	3189.0627	14.2220	13.9273	198.0748
90	441.1030	0.0023	0.0002	0.0702	6287.1854	14.2533	14.0812	200.7042
100	867.7163	0.0012	0.0001	0.0701	12381.6618	14.2693	14.1703	202.2001
110	1706.9293	0.0006	0.0000	0.0700	24370.4193	14.2773	14.2212	203.0415
120	3357.7884	0.0003	0.0000	0.0700	47954.1198	14.2815	14.2500	203.5103

Interest Rate	8.00%

	single payment		Uniform payment series				Arithmetic Gradient	
n	F/P	P/F	A/F	A/P	F/A	P/A	A/G	P/G
1	1.0800	0.9259	1.0000	1.0800	1.0000	0.9259	0.0000	0.0000
2	1.1664	0.8573	0.4808	0.5608	2.0800	1.7833	0.4808	0.8573
3	1.2597	0.7938	0.3080	0.3880	3.2464	2.5771	0.9487	2.4450
4	1.3605	0.7350	0.2219	0.3019	4.5061	3.3121	1.4040	4.6501
5	1.4693	0.6806	0.1705	0.2505	5.8666	3.9927	1.8465	7.3724
6	1.5869	0.6302	0.1363	0.2163	7.3359	4.6229	2.2763	10.5233
7	1.7138	0.5835	0.1121	0.1921	8.9228	5.2064	2.6937	14.0242
8	1.8509	0.5403	0.0940	0.1740	10.6366	5.7466	3.0985	17.8061
9	1.9990	0.5002	0.0801	0.1601	12.4876	6.2469	3.4910	21.8081
10	2.1589	0.4632	0.0690	0.1490	14.4866	6.7101	3.8713	25.9768
11	2.3316	0.4289	0.0601	0.1401	16.6455	7.1390	4.2395	30.2657
12	2.5182	0.3971	0.0527	0.1327	18.9771	7.5361	4.5957	34.6339
13	2.7196	0.3677	0.0465	0.1265	21.4953	7.9038	4.9402	39.0463
14	2.9372	0.3405	0.0413	0.1213	24.2149	8.2442	5.2731	43.4723
15	3.1722	0.3152	0.0368	0.1168	27.1521	8.5595	5.5945	47.8857
16	3.4259	0.2919	0.0330	0.1130	30.3243	8.8514	5.9046	52.2640
17	3.7000	0.2703	0.0296	0.1096	33.7502	9.1216	6.2037	56.5883
19	4.3157	0.2317	0.0241	0.1041	41.4463	9.6036	6.7697	65.0134
20	4.6610	0.2145	0.0219	0.1019	45.7620	9.8181	7.0369	69.0898
21	5.0338	0.1987	0.0198	0.0998	50.4229	10.0168	7.2940	73.0629
23	5.8715	0.1703	0.0164	0.0964	60.8933	10.3711	7.7786	80.6726
24	6.3412	0.1577	0.0150	0.0950	66.7648	10.5288	8.0066	84.2997
25	6.8485	0.1460	0.0137	0.0937	73.1059	10.6748	8.2254	87.8041
27	7.9881	0.1252	0.0114	0.0914	87.3508	10.9352	8.6363	94.4390
28	8.6271	0.1159	0.0105	0.0905	95.3388	11.0511	8.8289	97.5687
29	9.3173	0.1073	0.0096	0.0896	103.9659	11.1584	9.0133	100.5738
30	10.0627	0.0994	0.0088	0.0888	113.2832	11.2578	9.1897	103.4558
35	14.7853	0.0676	0.0058	0.0858	172.3168	11.6546	9.9611	116.0920
36	15.9682	0.0626	0.0053	0.0853	187.1021	11.7172	10.0949	118.2839
40	21.7245	0.0460	0.0039	0.0839	259.0565	11.9246	10.5699	126.0422
45	31.9204	0.0313	0.0026	0.0826	386.5056	12.1084	11.0447	133.7331
50	46.9016	0.0213	0.0017	0.0817	573.7702	12.2335	11.4107	139.5928
60	101.2571	0.0099	0.0008	0.0808	1253.2133	12.3766	11.9015	147.3000
65	148.7798	0.0067	0.0005	0.0805	1847.2481	12.4160	12.0602	149.7387
70	218.6064	0.0046	0.0004	0.0804	2720.0801	12.4428	12.1783	151.5326
80	471.9548	0.0021	0.0002	0.0802	5886.9354	12.4735	12.3301	153.8001
90	1018.9151	0.0010	0.0001	0.0801	12723.9386	12.4877	12.4116	154.9925
100	2199.7613	0.0005	0.0000	0.0800	27484.5157	12.4943	12.4545	155.6107
110	4749.1196	0.0002	0.0000	0.0800	59351.4946	12.4974	12.4768	155.9276
120	10252.9929	0.0001	0.0000	0.0800	128149.9118	12.4988	12.4883	156.0885

Interest Rate	9.00%

	single payment		Uniform payment series				Arithmetic Gradient	
n	F/P	P/F	A/F	A/P	F/A	P/A	A/G	P/G
1	1.0900	0.9174	1.0000	1.0900	1.0000	0.9174	0.0000	0.0000
2	1.1881	0.8417	0.4785	0.5685	2.0900	1.7591	0.4785	0.8417
3	1.2950	0.7722	0.3051	0.3951	3.2781	2.5313	0.9426	2.3860
4	1.4116	0.7084	0.2187	0.3087	4.5731	3.2397	1.3925	4.5113
5	1.5386	0.6499	0.1671	0.2571	5.9847	3.8897	1.8282	7.1110
6	1.6771	0.5963	0.1329	0.2229	7.5233	4.4859	2.2498	10.0924
7	1.8280	0.5470	0.1087	0.1987	9.2004	5.0330	2.6574	13.3746
8	1.9926	0.5019	0.0907	0.1807	11.0285	5.5348	3.0512	16.8877
9	2.1719	0.4604	0.0768	0.1668	13.0210	5.9952	3.4312	20.5711
10	2.3674	0.4224	0.0658	0.1558	15.1929	6.4177	3.7978	24.3728
11	2.5804	0.3875	0.0569	0.1469	17.5603	6.8052	4.1510	28.2481
12	2.8127	0.3555	0.0497	0.1397	20.1407	7.1607	4.4910	32.1590
13	3.0658	0.3262	0.0436	0.1336	22.9534	7.4869	4.8182	36.0731
14	3.3417	0.2992	0.0384	0.1284	26.0192	7.7862	5.1326	39.9633
15	3.6425	0.2745	0.0341	0.1241	29.3609	8.0607	5.4346	43.8069
16	3.9703	0.2519	0.0303	0.1203	33.0034	8.3126	5.7245	47.5849
17	4.3276	0.2311	0.0270	0.1170	36.9737	8.5436	6.0024	51.2821
19	5.1417	0.1945	0.0217	0.1117	46.0185	8.9501	6.5236	58.3868
20	5.6044	0.1784	0.0195	0.1095	51.1601	9.1285	6.7674	61.7770
21	6.1088	0.1637	0.0176	0.1076	56.7645	9.2922	7.0006	65.0509
23	7.2579	0.1378	0.0144	0.1044	69.5319	9.5802	7.4357	71.2359
24	7.9111	0.1264	0.0130	0.1030	76.7898	9.7066	7.6384	74.1433
25	8.6231	0.1160	0.0118	0.1018	84.7009	9.8226	7.8316	76.9265
27	10.2451	0.0976	0.0097	0.0997	102.7231	10.0266	8.1906	82.1241
28	11.1671	0.0895	0.0089	0.0989	112.9682	10.1161	8.3571	84.5419
29	12.1722	0.0822	0.0081	0.0981	124.1354	10.1983	8.5154	86.8422
30	13.2677	0.0754	0.0073	0.0973	136.3075	10.2737	8.6657	89.0280
35	20.4140	0.0490	0.0046	0.0946	215.7108	10.5668	9.3083	98.3590
36	22.2512	0.0449	0.0042	0.0942	236.1247	10.6118	9.4171	99.9319
40	31.4094	0.0318	0.0030	0.0930	337.8824	10.7574	9.7957	105.3762
45	48.3273	0.0207	0.0019	0.0919	525.8587	10.8812	10.1603	110.5561
50	74.3575	0.0134	0.0012	0.0912	815.0836	10.9617	10.4295	114.3251
60	176.0313	0.0057	0.0005	0.0905	1944.7921	11.0480	10.7683	118.9683
65	270.8460	0.0037	0.0003	0.0903	2998.2885	11.0701	10.8702	120.3344
70	416.7301	0.0024	0.0002	0.0902	4619.2232	11.0844	10.9427	121.2942
80	986.5517	0.0010	0.0001	0.0901	10950.5741	11.0998	11.0299	122.4306
90	2335.5266	0.0004	0.0000	0.0900	25939.1842	11.1064	11.0726	122.9758
100	5529.0408	0.0002	0.0000	0.0900	61422.6755	11.1091	11.0930	123.2335
110	13089.2503	0.0001	0.0000	0.0900	145425.0036	11.1103	11.1027	123.3540
120	30987.0157	0.0000	0.0000	0.0900	344289.0639	11.1108	11.1072	123.4098

Interest Rate	10.00%

	single payment		Uniform payment series				Arithmetic Gradient	
n	F/P	P/F	A/F	A/P	F/A	P/A	A/G	P/G
1	1.1000	0.9091	1.0000	1.1000	1.0000	0.9091	0.0000	0.0000
2	1.2100	0.8264	0.4762	0.5762	2.1000	1.7355	0.4762	0.8264
3	1.3310	0.7513	0.3021	0.4021	3.3100	2.4869	0.9366	2.3291
4	1.4641	0.6830	0.2155	0.3155	4.6410	3.1699	1.3812	4.3781
5	1.6105	0.6209	0.1638	0.2638	6.1051	3.7908	1.8101	6.8618
6	1.7716	0.5645	0.1296	0.2296	7.7156	4.3553	2.2236	9.6842
7	1.9487	0.5132	0.1054	0.2054	9.4872	4.8684	2.6216	12.7631
8	2.1436	0.4665	0.0874	0.1874	11.4359	5.3349	3.0045	16.0287
9	2.3579	0.4241	0.0736	0.1736	13.5795	5.7590	3.3724	19.4215
10	2.5937	0.3855	0.0627	0.1627	15.9374	6.1446	3.7255	22.8913
11	2.8531	0.3505	0.0540	0.1540	18.5312	6.4951	4.0641	26.3963
12	3.1384	0.3186	0.0468	0.1468	21.3843	6.8137	4.3884	29.9012
13	3.4523	0.2897	0.0408	0.1408	24.5227	7.1034	4.6988	33.3772
14	3.7975	0.2633	0.0357	0.1357	27.9750	7.3667	4.9955	36.8005
15	4.1772	0.2394	0.0315	0.1315	31.7725	7.6061	5.2789	40.1520
16	4.5950	0.2176	0.0278	0.1278	35.9497	7.8237	5.5493	43.4164
17	5.0545	0.1978	0.0247	0.1247	40.5447	8.0216	5.8071	46.5819
19	6.1159	0.1635	0.0195	0.1195	51.1591	8.3649	6.2861	52.5827
20	6.7275	0.1486	0.0175	0.1175	57.2750	8.5136	6.5081	55.4069
21	7.4002	0.1351	0.0156	0.1156	64.0025	8.6487	6.7189	58.1095
23	8.9543	0.1117	0.0126	0.1126	79.5430	8.8832	7.1085	63.1462
24	9.8497	0.1015	0.0113	0.1113	88.4973	8.9847	7.2881	65.4813
25	10.8347	0.0923	0.0102	0.1102	98.3471	9.0770	7.4580	67.6964
27	13.1100	0.0763	0.0083	0.1083	121.0999	9.2372	7.7704	71.7773
28	14.4210	0.0693	0.0075	0.1075	134.2099	9.3066	7.9137	73.6495
29	15.8631	0.0630	0.0067	0.1067	148.6309	9.3696	8.0489	75.4146
30	17.4494	0.0573	0.0061	0.1061	164.4940	9.4269	8.1762	77.0766
35	28.1024	0.0356	0.0037	0.1037	271.0244	9.6442	8.7086	83.9872
36	30.9127	0.0323	0.0033	0.1033	299.1268	9.6765	8.7965	85.1194
40	45.2593	0.0221	0.0023	0.1023	442.5926	9.7791	9.0962	88.9525
45	72.8905	0.0137	0.0014	0.1014	718.9048	9.8628	9.3740	92.4544
50	117.3909	0.0085	0.0009	0.1009	1163.9085	9.9148	9.5704	94.8889
60	304.4816	0.0033	0.0003	0.1003	3034.8164	9.9672	9.8023	97.7010
65	490.3707	0.0020	0.0002	0.1002	4893.7073	9.9796	9.8672	98.4705
70	789.7470	0.0013	0.0001	0.1001	7887.4696	9.9873	9.9113	98.9870
80	2048.4002	0.0005	0.0000	0.1000	20474.0021	9.9951	9.9609	99.5606
90	5313.0226	0.0002	0.0000	0.1000	53120.2261	9.9981	9.9831	99.8118
100	13780.6123	0.0001	0.0000	0.1000	137796.1234	9.9993	9.9927	99.9202
110	35743.3594	0.0000	0.0000	0.1000	357423.5935	9.9997	9.9969	99.9664
120	92709.0688	0.0000	0.0000	0.1000	927080.6882	9.9999	9.9987	99.9860

Interest Rate	11.00%

	single payment		Uniform payment series				Arithmetic Gradient	
n	F/P	P/F	A/F	A/P	F/A	P/A	A/G	P/G
1	1.1100	0.9009	1.0000	1.1100	1.0000	0.9009	0.0000	0.0000
2	1.2321	0.8116	0.4739	0.5839	2.1100	1.7125	0.4739	0.8116
3	1.3676	0.7312	0.2992	0.4092	3.3421	2.4437	0.9306	2.2740
4	1.5181	0.6587	0.2123	0.3223	4.7097	3.1024	1.3700	4.2502
5	1.6851	0.5935	0.1606	0.2706	6.2278	3.6959	1.7923	6.6240
6	1.8704	0.5346	0.1264	0.2364	7.9129	4.2305	2.1976	9.2972
7	2.0762	0.4817	0.1022	0.2122	9.7833	4.7122	2.5863	12.1872
8	2.3045	0.4339	0.0843	0.1943	11.8594	5.1461	2.9585	15.2246
9	2.5580	0.3909	0.0706	0.1806	14.1640	5.5370	3.3144	18.3520
10	2.8394	0.3522	0.0598	0.1698	16.7220	5.8892	3.6544	21.5217
11	3.1518	0.3173	0.0511	0.1611	19.5614	6.2065	3.9788	24.6945
12	3.4985	0.2858	0.0440	0.1540	22.7132	6.4924	4.2879	27.8388
13	3.8833	0.2575	0.0382	0.1482	26.2116	6.7499	4.5822	30.9290
14	4.3104	0.2320	0.0332	0.1432	30.0949	6.9819	4.8619	33.9449
15	4.7846	0.2090	0.0291	0.1391	34.4054	7.1909	5.1275	36.8709
16	5.3109	0.1883	0.0255	0.1355	39.1899	7.3792	5.3794	39.6953
17	5.8951	0.1696	0.0225	0.1325	44.5008	7.5488	5.6180	42.4095
19	7.2633	0.1377	0.0176	0.1276	56.9395	7.8393	6.0574	47.4856
20	8.0623	0.1240	0.0156	0.1256	64.2028	7.9633	6.2590	49.8423
21	8.9492	0.1117	0.0138	0.1238	72.2651	8.0751	6.4491	52.0771
23	11.0263	0.0907	0.0110	0.1210	91.1479	8.2664	6.7969	56.1864
24	12.2392	0.0817	0.0098	0.1198	102.1742	8.3481	6.9555	58.0656
25	13.5855	0.0736	0.0087	0.1187	114.4133	8.4217	7.1045	59.8322
27	16.7386	0.0597	0.0070	0.1170	143.0786	8.5478	7.3754	63.0433
28	18.5799	0.0538	0.0063	0.1163	159.8173	8.6016	7.4982	64.4965
29	20.6237	0.0485	0.0056	0.1156	178.3972	8.6501	7.6131	65.8542
30	22.8923	0.0437	0.0050	0.1150	199.0209	8.6938	7.7206	67.1210
35	38.5749	0.0259	0.0029	0.1129	341.5896	8.8552	8.1594	72.2538
36	42.8181	0.0234	0.0026	0.1126	380.1644	8.8786	8.2300	73.0712
40	65.0009	0.0154	0.0017	0.1117	581.8261	8.9511	8.4659	75.7789
45	109.5302	0.0091	0.0010	0.1110	986.6386	9.0079	8.6763	78.1551
50	184.5648	0.0054	0.0006	0.1106	1668.7712	9.0417	8.8185	79.7341
60	524.0572	0.0019	0.0002	0.1102	4755.0658	9.0736	8.9762	81.4461
65	883.0669	0.0011	0.0001	0.1101	8018.7903	9.0806	9.0172	81.8819
70	1488.0191	0.0007	0.0001	0.1101	13518.3557	9.0848	9.0438	82.1614
80	4225.1128	0.0002	0.0000	0.1100	38401.0250	9.0888	9.0720	82.4529
90	11996.8738	0.0001	0.0000	0.1100	109053.3983	9.0902	9.0834	82.5695
100	34064.1753	0.0000	0.0000	0.1100	309665.2297	9.0906	9.0880	82.6155
110	96722.5341	0.0000	0.0000	0.1100	879286.6739	9.0908	9.0898	82.6334
120	274635.9932	0.0000	0.0000	0.1100	2496681.7568	9.0909	9.0905	82.6404

Interest Rate	12.00%

	single payment		Uniform payment series				Arithmetic Gradient	
n	F/P	P/F	A/F	A/P	F/A	P/A	A/G	P/G
1	1.1200	0.8929	1.0000	1.1200	1.0000	0.8929	0.0000	0.0000
2	1.2544	0.7972	0.4717	0.5917	2.1200	1.6901	0.4717	0.7972
3	1.4049	0.7118	0.2963	0.4163	3.3744	2.4018	0.9246	2.2208
4	1.5735	0.6355	0.2092	0.3292	4.7793	3.0373	1.3589	4.1273
5	1.7623	0.5674	0.1574	0.2774	6.3528	3.6048	1.7746	6.3970
6	1.9738	0.5066	0.1232	0.2432	8.1152	4.1114	2.1720	8.9302
7	2.2107	0.4523	0.0991	0.2191	10.0890	4.5638	2.5515	11.6443
8	2.4760	0.4039	0.0813	0.2013	12.2997	4.9676	2.9131	14.4714
9	2.7731	0.3606	0.0677	0.1877	14.7757	5.3282	3.2574	17.3563
10	3.1058	0.3220	0.0570	0.1770	17.5487	5.6502	3.5847	20.2541
11	3.4785	0.2875	0.0484	0.1684	20.6546	5.9377	3.8953	23.1288
12	3.8960	0.2567	0.0414	0.1614	24.1331	6.1944	4.1897	25.9523
13	4.3635	0.2292	0.0357	0.1557	28.0291	6.4235	4.4683	28.7024
14	4.8871	0.2046	0.0309	0.1509	32.3926	6.6282	4.7317	31.3624
15	5.4736	0.1827	0.0268	0.1468	37.2797	6.8109	4.9803	33.9202
16	6.1304	0.1631	0.0234	0.1434	42.7533	6.9740	5.2147	36.3670
17	6.8660	0.1456	0.0205	0.1405	48.8837	7.1196	5.4353	38.6973
19	8.6128	0.1161	0.0158	0.1358	63.4397	7.3658	5.8375	42.9979
20	9.6463	0.1037	0.0139	0.1339	72.0524	7.4694	6.0202	44.9676
21	10.8038	0.0926	0.0122	0.1322	81.6987	7.5620	6.1913	46.8188
23	13.5523	0.0738	0.0096	0.1296	104.6029	7.7184	6.5010	50.1776
24	15.1786	0.0659	0.0085	0.1285	118.1552	7.7843	6.6406	51.6929
25	17.0001	0.0588	0.0075	0.1275	133.3339	7.8431	6.7708	53.1046
27	21.3249	0.0469	0.0059	0.1259	169.3740	7.9426	7.0049	55.6369
28	23.8839	0.0419	0.0052	0.1252	190.6989	7.9844	7.1098	56.7674
29	26.7499	0.0374	0.0047	0.1247	214.5828	8.0218	7.2071	57.8141
30	29.9599	0.0334	0.0041	0.1241	241.3327	8.0552	7.2974	58.7821
35	52.7996	0.0189	0.0023	0.1223	431.6635	8.1755	7.6577	62.6052
36	59.1356	0.0169	0.0021	0.1221	484.4631	8.1924	7.7141	63.1970
40	93.0510	0.0107	0.0013	0.1213	767.0914	8.2438	7.8988	65.1159
45	163.9876	0.0061	0.0007	0.1207	1358.2300	8.2825	8.0572	66.7342
50	289.0022	0.0035	0.0004	0.1204	2400.0182	8.3045	8.1597	67.7624
60	897.5969	0.0011	0.0001	0.1201	7471.6411	8.3240	8.2664	68.8100
65	1581.8725	0.0006	0.0001	0.1201	13173.9374	8.3281	8.2922	69.0581
70	2787.7998	0.0004	0.0000	0.1200	23223.3319	8.3303	8.3082	69.2103
80	8658.4831	0.0001	0.0000	0.1200	72145.6925	8.3324	8.3241	69.3594
90	26891.9342	0.0000	0.0000	0.1200	224091.1185	8.3330	8.3300	69.4140
100	83522.2657	0.0000	0.0000	0.1200	696010.5477	8.3332	8.3321	69.4336
110	259407.4794	0.0000	0.0000	0.1200	2161720.6614	8.3333	8.3329	69.4406
120	805680.2550	0.0000	0.0000	0.1200	6713993.7918	8.3333	8.3332	69.4431

Interest Rate	15.00%

	single payment		Uniform payment series				Arithmetic Gradient	
n	F/P	P/F	A/F	A/P	F/A	P/A	A/G	P/G
1	1.1500	0.8696	1.0000	1.1500	1.0000	0.8696	0.0000	0.0000
2	1.3225	0.7561	0.4651	0.6151	2.1500	1.6257	0.4651	0.7561
3	1.5209	0.6575	0.2880	0.4380	3.4725	2.2832	0.9071	2.0712
4	1.7490	0.5718	0.2003	0.3503	4.9934	2.8550	1.3263	3.7864
5	2.0114	0.4972	0.1483	0.2983	6.7424	3.3522	1.7228	5.7751
6	2.3131	0.4323	0.1142	0.2642	8.7537	3.7845	2.0972	7.9368
7	2.6600	0.3759	0.0904	0.2404	11.0668	4.1604	2.4498	10.1924
8	3.0590	0.3269	0.0729	0.2229	13.7268	4.4873	2.7813	12.4807
9	3.5179	0.2843	0.0596	0.2096	16.7858	4.7716	3.0922	14.7548
10	4.0456	0.2472	0.0493	0.1993	20.3037	5.0188	3.3832	16.9795
11	4.6524	0.2149	0.0411	0.1911	24.3493	5.2337	3.6549	19.1289
12	5.3503	0.1869	0.0345	0.1845	29.0017	5.4206	3.9082	21.1849
13	6.1528	0.1625	0.0291	0.1791	34.3519	5.5831	4.1438	23.1352
14	7.0757	0.1413	0.0247	0.1747	40.5047	5.7245	4.3624	24.9725
15	8.1371	0.1229	0.0210	0.1710	47.5804	5.8474	4.5650	26.6930
16	9.3576	0.1069	0.0179	0.1679	55.7175	5.9542	4.7522	28.2960
17	10.7613	0.0929	0.0154	0.1654	65.0751	6.0472	4.9251	29.7828
19	14.2318	0.0703	0.0113	0.1613	88.2118	6.1982	5.2307	32.4213
20	16.3665	0.0611	0.0098	0.1598	102.4436	6.2593	5.3651	33.5822
21	18.8215	0.0531	0.0084	0.1584	118.8101	6.3125	5.4883	34.6448
23	24.8915	0.0402	0.0063	0.1563	159.2764	6.3988	5.7040	36.4988
24	28.6252	0.0349	0.0054	0.1554	184.1678	6.4338	5.7979	37.3023
25	32.9190	0.0304	0.0047	0.1547	212.7930	6.4641	5.8834	38.0314
27	43.5353	0.0230	0.0035	0.1535	283.5688	6.5135	6.0319	39.2890
28	50.0656	0.0200	0.0031	0.1531	327.1041	6.5335	6.0960	39.8283
29	57.5755	0.0174	0.0027	0.1527	377.1697	6.5509	6.1541	40.3146
30	66.2118	0.0151	0.0023	0.1523	434.7451	6.5660	6.2066	40.7526
35	133.1755	0.0075	0.0011	0.1511	881.1702	6.6166	6.4019	42.3586
36	153.1519	0.0065	0.0010	0.1510	1014.3457	6.6231	6.4301	42.5872
40	267.8635	0.0037	0.0006	0.1506	1779.0903	6.6418	6.5168	43.2830
45	538.7693	0.0019	0.0003	0.1503	3585.1285	6.6543	6.5830	43.8051

Interest Rate	17.00%

	single payment		Uniform payment series				Arithmetic Gradient	
n	F/P	P/F	A/F	A/P	F/A	P/A	A/G	P/G
1	1.1700	0.8547	1.0000	1.1700	1.0000	0.8547	0.0000	0.0000
2	1.3689	0.7305	0.4608	0.6308	2.1700	1.5852	0.4608	0.7305
3	1.6016	0.6244	0.2826	0.4526	3.5389	2.2096	0.8958	1.9793
4	1.8739	0.5337	0.1945	0.3645	5.1405	2.7432	1.3051	3.5802
5	2.1924	0.4561	0.1426	0.3126	7.0144	3.1993	1.6893	5.4046
6	2.5652	0.3898	0.1086	0.2786	9.2068	3.5892	2.0489	7.3538
7	3.0012	0.3332	0.0849	0.2549	11.7720	3.9224	2.3845	9.3530
8	3.5115	0.2848	0.0677	0.2377	14.7733	4.2072	2.6969	11.3465
9	4.1084	0.2434	0.0547	0.2247	18.2847	4.4506	2.9870	13.2937
10	4.8068	0.2080	0.0447	0.2147	22.3931	4.6586	3.2555	15.1661
11	5.6240	0.1778	0.0368	0.2068	27.1999	4.8364	3.5035	16.9442
12	6.5801	0.1520	0.0305	0.2005	32.8239	4.9884	3.7318	18.6159
13	7.6987	0.1299	0.0254	0.1954	39.4040	5.1183	3.9417	20.1746
14	9.0075	0.1110	0.0212	0.1912	47.1027	5.2293	4.1340	21.6178
15	10.5387	0.0949	0.0178	0.1878	56.1101	5.3242	4.3098	22.9463
16	12.3303	0.0811	0.0150	0.1850	66.6488	5.4053	4.4702	24.1628
17	14.4265	0.0693	0.0127	0.1827	78.9792	5.4746	4.6162	25.2719
19	19.7484	0.0506	0.0091	0.1791	110.2846	5.5845	4.8689	27.1905
20	23.1056	0.0433	0.0077	0.1777	130.0329	5.6278	4.9776	28.0128
21	27.0336	0.0370	0.0065	0.1765	153.1385	5.6648	5.0757	28.7526
23	37.0062	0.0270	0.0047	0.1747	211.8013	5.7234	5.2436	30.0111
24	43.2973	0.0231	0.0040	0.1740	248.8076	5.7465	5.3149	30.5423
25	50.6578	0.0197	0.0034	0.1734	292.1049	5.7662	5.3789	31.0160
27	69.3455	0.0144	0.0025	0.1725	402.0323	5.7975	5.4873	31.8128
28	81.1342	0.0123	0.0021	0.1721	471.3778	5.8099	5.5329	32.1456
29	94.9271	0.0105	0.0018	0.1718	552.5121	5.8204	5.5736	32.4405
30	111.0647	0.0090	0.0015	0.1715	647.4391	5.8294	5.6098	32.7016
35	243.5035	0.0041	0.0007	0.1707	1426.4910	5.8582	5.7380	33.6145
36	284.8991	0.0035	0.0006	0.1706	1669.9945	5.8617	5.7555	33.7373
40	533.8687	0.0019	0.0003	0.1703	3134.5218	5.8713	5.8073	34.0965
45	1170.4794	0.0009	0.0001	0.1701	6879.2907	5.8773	5.8439	34.3464
50	2566.2153	0.0004	0.0001	0.1701	#########	5.8801	5.8629	34.4740
60	#########	0.0001	0.0000	0.1700	#########	5.8819	5.8775	34.5707
65	#########	0.0000	0.0000	0.1700	#########	5.8821	5.8799	34.5867
70	#########	0.0000	0.0000	0.1700	#########	5.8823	5.8812	34.5945
80	#########	0.0000	0.0000	0.1700	#########	5.8823	5.8821	34.6003
90	#########	0.0000	0.0000	0.1700	#########	5.8823	5.8823	34.6017
100	#########	0.0000	0.0000	0.1700	#########	5.8824	5.8823	34.6020
110	#########	0.0000	0.0000	0.1700	#########	5.8824	5.8823	34.6021
120	#########	0.0000	0.0000	0.1700	#########	5.8824	5.8824	34.6021

Interest Rate	20.00%

n	single payment		Uniform payment series				Arithmetic Gradient	
	F/P	P/F	A/F	A/P	F/A	P/A	A/G	P/G
1	1.2000	0.8333	1.0000	1.2000	1.0000	0.8333	0.0000	0.0000
2	1.4400	0.6944	0.4545	0.6545	2.2000	1.5278	0.4545	0.6944
3	1.7280	0.5787	0.2747	0.4747	3.6400	2.1065	0.8791	1.8519
4	2.0736	0.4823	0.1863	0.3863	5.3680	2.5887	1.2742	3.2986
5	2.4883	0.4019	0.1344	0.3344	7.4416	2.9906	1.6405	4.9061
6	2.9860	0.3349	0.1007	0.3007	9.9299	3.3255	1.9788	6.5806
7	3.5832	0.2791	0.0774	0.2774	12.9159	3.6046	2.2902	8.2551
8	4.2998	0.2326	0.0606	0.2606	16.4991	3.8372	2.5756	9.8831
9	5.1598	0.1938	0.0481	0.2481	20.7989	4.0310	2.8364	11.4335
10	6.1917	0.1615	0.0385	0.2385	25.9587	4.1925	3.0739	12.8871
11	7.4301	0.1346	0.0311	0.2311	32.1504	4.3271	3.2893	14.2330
12	8.9161	0.1122	0.0253	0.2253	39.5805	4.4392	3.4841	15.4667
13	10.6993	0.0935	0.0206	0.2206	48.4966	4.5327	3.6597	16.5883
14	12.8392	0.0779	0.0169	0.2169	59.1959	4.6106	3.8175	17.6008
15	15.4070	0.0649	0.0139	0.2139	72.0351	4.6755	3.9588	18.5095
16	18.4884	0.0541	0.0114	0.2114	87.4421	4.7296	4.0851	19.3208
17	22.1861	0.0451	0.0094	0.2094	105.9306	4.7746	4.1976	20.0419
19	31.9480	0.0313	0.0065	0.2065	154.7400	4.8435	4.3861	21.2439
20	38.3376	0.0261	0.0054	0.2054	186.6880	4.8696	4.4643	21.7395
21	46.0051	0.0217	0.0044	0.2044	225.0256	4.8913	4.5334	22.1742
23	66.2474	0.0151	0.0031	0.2031	326.2369	4.9245	4.6475	22.8867
24	79.4968	0.0126	0.0025	0.2025	392.4842	4.9371	4.6943	23.1760
25	95.3962	0.0105	0.0021	0.2021	471.9811	4.9476	4.7352	23.4276
27	137.3706	0.0073	0.0015	0.2015	681.8528	4.9636	4.8020	23.8353
28	164.8447	0.0061	0.0012	0.2012	819.2233	4.9697	4.8291	23.9991
29	197.8136	0.0051	0.0010	0.2010	984.0680	4.9747	4.8527	24.1406
30	237.3763	0.0042	0.0008	0.2008	1181.8816	4.9789	4.8731	24.2628
35	590.6682	0.0017	0.0003	0.2003	2948.3411	4.9915	4.9406	24.6614
36	708.8019	0.0014	0.0003	0.2003	3539.0094	4.9929	4.9491	24.7108
40	1469.7716	0.0007	0.0001	0.2001	7343.8578	4.9966	4.9728	24.8469
45	3657.2620	0.0003	0.0001	0.2001	#########	4.9986	4.9877	24.9316
50	9100.4382	0.0001	0.0000	0.2000	#########	4.9995	4.9945	24.9698

Interest Rate	25.00%

	single payment		Uniform payment series				Arithmetic Gradient	
n	F/P	P/F	A/F	A/P	F/A	P/A	A/G	P/G
1	1.2500	0.8000	1.0000	1.2500	1.0000	0.8000	0.0000	0.0000
2	1.5625	0.6400	0.4444	0.6944	2.2500	1.4400	0.4444	0.6400
3	1.9531	0.5120	0.2623	0.5123	3.8125	1.9520	0.8525	1.6640
4	2.4414	0.4096	0.1734	0.4234	5.7656	2.3616	1.2249	2.8928
5	3.0518	0.3277	0.1218	0.3718	8.2070	2.6893	1.5631	4.2035
6	3.8147	0.2621	0.0888	0.3388	11.2588	2.9514	1.8683	5.5142
7	4.7684	0.2097	0.0663	0.3163	15.0735	3.1611	2.1424	6.7725
8	5.9605	0.1678	0.0504	0.3004	19.8419	3.3289	2.3872	7.9469
9	7.4506	0.1342	0.0388	0.2888	25.8023	3.4631	2.6048	9.0207
10	9.3132	0.1074	0.0301	0.2801	33.2529	3.5705	2.7971	9.9870
11	11.6415	0.0859	0.0235	0.2735	42.5661	3.6564	2.9663	10.8460
12	14.5519	0.0687	0.0184	0.2684	54.2077	3.7251	3.1145	11.6020
13	18.1899	0.0550	0.0145	0.2645	68.7596	3.7801	3.2437	12.2617
14	22.7374	0.0440	0.0115	0.2615	86.9495	3.8241	3.3559	12.8334
15	28.4217	0.0352	0.0091	0.2591	109.6868	3.8593	3.4530	13.3260
16	35.5271	0.0281	0.0072	0.2572	138.1085	3.8874	3.5366	13.7482
17	44.4089	0.0225	0.0058	0.2558	173.6357	3.9099	3.6084	14.1085
19	69.3889	0.0144	0.0037	0.2537	273.5558	3.9424	3.7222	14.6741
20	86.7362	0.0115	0.0029	0.2529	342.9447	3.9539	3.7667	14.8932
21	108.4202	0.0092	0.0023	0.2523	429.6809	3.9631	3.8045	15.0777
23	169.4066	0.0059	0.0015	0.2515	673.6264	3.9764	3.8634	15.3625
24	211.7582	0.0047	0.0012	0.2512	843.0329	3.9811	3.8861	15.4711
25	264.6978	0.0038	0.0009	0.2509	1054.7912	3.9849	3.9052	15.5618
27	413.5903	0.0024	0.0006	0.2506	1650.3612	3.9903	3.9346	15.7002
28	516.9879	0.0019	0.0005	0.2505	2063.9515	3.9923	3.9457	15.7524
29	646.2349	0.0015	0.0004	0.2504	2580.9394	3.9938	3.9551	15.7957
30	807.7936	0.0012	0.0003	0.2503	3227.1743	3.9950	3.9628	15.8316
35	2465.1903	0.0004	0.0001	0.2501	9856.7613	3.9984	3.9858	15.9367
36	3081.4879	0.0003	0.0001	0.2501	#########	3.9987	3.9883	15.9481
40	7523.1638	0.0001	0.0000	0.2500	#########	3.9995	3.9947	15.9766

Interest Rate	30.00%

	single payment		Uniform payment series				Arithmetic Gradient	
n	F/P	P/F	A/F	A/P	F/A	P/A	A/G	P/G
1	1.3000	0.7692	1.0000	1.3000	1.0000	0.7692	0.0000	0.0000
2	1.6900	0.5917	0.4348	0.7348	2.3000	1.3609	0.4348	0.5917
3	2.1970	0.4552	0.2506	0.5506	3.9900	1.8161	0.8271	1.5020
4	2.8561	0.3501	0.1616	0.4616	6.1870	2.1662	1.1783	2.5524
5	3.7129	0.2693	0.1106	0.4106	9.0431	2.4356	1.4903	3.6297
6	4.8268	0.2072	0.0784	0.3784	12.7560	2.6427	1.7654	4.6656
7	6.2749	0.1594	0.0569	0.3569	17.5828	2.8021	2.0063	5.6218
8	8.1573	0.1226	0.0419	0.3419	23.8577	2.9247	2.2156	6.4800
9	10.6045	0.0943	0.0312	0.3312	32.0150	3.0190	2.3963	7.2343
10	13.7858	0.0725	0.0235	0.3235	42.6195	3.0915	2.5512	7.8872
11	17.9216	0.0558	0.0177	0.3177	56.4053	3.1473	2.6833	8.4452
12	23.2981	0.0429	0.0135	0.3135	74.3270	3.1903	2.7952	8.9173
13	30.2875	0.0330	0.0102	0.3102	97.6250	3.2233	2.8895	9.3135
14	39.3738	0.0254	0.0078	0.3078	127.9125	3.2487	2.9685	9.6437
15	51.1859	0.0195	0.0060	0.3060	167.2863	3.2682	3.0344	9.9172
16	66.5417	0.0150	0.0046	0.3046	218.4722	3.2832	3.0892	10.1426
17	86.5042	0.0116	0.0035	0.3035	285.0139	3.2948	3.1345	10.3276
19	146.1920	0.0068	0.0021	0.3021	483.9734	3.3105	3.2025	10.6019
20	190.0496	0.0053	0.0016	0.3016	630.1655	3.3158	3.2275	10.7019
21	247.0645	0.0040	0.0012	0.3012	820.2151	3.3198	3.2480	10.7828
23	417.5391	0.0024	0.0007	0.3007	1388.4635	3.3254	3.2781	10.9009
24	542.8008	0.0018	0.0006	0.3006	1806.0026	3.3272	3.2890	10.9433
25	705.6410	0.0014	0.0004	0.3004	2348.8033	3.3286	3.2979	10.9773
27	1192.5333	0.0008	0.0003	0.3003	3971.7776	3.3305	3.3107	11.0263
28	1550.2933	0.0006	0.0002	0.3002	5164.3109	3.3312	3.3153	11.0437
29	2015.3813	0.0005	0.0001	0.3001	6714.6042	3.3317	3.3189	11.0576
30	2619.9956	0.0004	0.0001	0.3001	8729.9855	3.3321	3.3219	11.0687
35	9727.8604	0.0001	0.0000	0.3000	32422.8681	3.3330	3.3297	11.0980
36	12646.2186	0.0001	0.0000	0.3000	42150.7285	3.3331	3.3305	11.1007
40	36118.8648	0.0000	0.0000	0.3000	120392.8827	3.3332	3.3322	11.1071

Interest Rate	35.00%

	single payment		Uniform payment series				rithmetic Gradie	
n	F/P	P/F	A/F	A/P	F/A	P/A	A/G	P/G
1	1.3500	0.7407	1.0000	1.3500	1.0000	0.7407	0.0000	0.0000
2	1.8225	0.5487	0.4255	0.7755	2.3500	1.2894	0.4255	0.5487
3	2.4604	0.4064	0.2397	0.5897	4.1725	1.6959	0.8029	1.3616
4	3.3215	0.3011	0.1508	0.5008	6.6329	1.9969	1.1341	2.2648
5	4.4840	0.2230	0.1005	0.4505	9.9544	2.2200	1.4220	3.1568
6	6.0534	0.1652	0.0693	0.4193	14.4384	2.3852	1.6698	3.9828
7	8.1722	0.1224	0.0488	0.3988	20.4919	2.5075	1.8811	4.7170
8	11.0324	0.0906	0.0349	0.3849	28.6640	2.5982	2.0597	5.3515
9	14.8937	0.0671	0.0252	0.3752	39.6964	2.6653	2.2094	5.8886
10	20.1066	0.0497	0.0183	0.3683	54.5902	2.7150	2.3338	6.3363
11	27.1439	0.0368	0.0134	0.3634	74.6967	2.7519	2.4364	6.7047
12	36.6442	0.0273	0.0098	0.3598	101.8406	2.7792	2.5205	7.0049
13	49.4697	0.0202	0.0072	0.3572	138.4848	2.7994	2.5889	7.2474
14	66.7841	0.0150	0.0053	0.3553	187.9544	2.8144	2.6443	7.4421
15	90.1585	0.0111	0.0039	0.3539	254.7385	2.8255	2.6889	7.5974
16	121.7139	0.0082	0.0029	0.3529	344.8970	2.8337	2.7246	7.7206
17	164.3138	0.0061	0.0021	0.3521	466.6109	2.8398	2.7530	7.8180
19	299.4619	0.0033	0.0012	0.3512	852.7483	2.8476	2.7935	7.9547
20	404.2736	0.0025	0.0009	0.3509	1152.2103	2.8501	2.8075	8.0017
21	545.7693	0.0018	0.0006	0.3506	1556.4838	2.8519	2.8186	8.0384
23	994.6646	0.0010	0.0004	0.3504	2839.0418	2.8543	2.8340	8.0890
24	1342.7973	0.0007	0.0003	0.3503	3833.7064	2.8550	2.8393	8.1061
25	1812.7763	0.0006	0.0002	0.3502	5176.5037	2.8556	2.8433	8.1194
27	3303.7848	0.0003	0.0001	0.3501	9436.5280	2.8563	2.8490	8.1374
28	4460.1095	0.0002	0.0001	0.3501	12740.3128	2.8565	2.8509	8.1435
29	6021.1478	0.0002	0.0001	0.3501	17200.4222	2.8567	2.8523	8.1481
30	8128.5495	0.0001	0.0000	0.3500	23221.5700	2.8568	2.8535	8.1517
35	36448.6878	0.0000	0.0000	0.3500	104136.2508	2.8571	2.8562	8.1603
36	49205.7285	0.0000	0.0000	0.3500	140584.9385	2.8571	2.8564	8.1610
40	163437.1347	0.0000	0.0000	0.3500	466960.3848	2.8571	2.8569	8.1625

Interest Rate	40.00%

	single payment		Uniform payment series				rithmetic Gradie	
n	F/P	P/F	A/F	A/P	F/A	P/A	A/G	P/G
1	1.4000	0.7143	1.0000	1.4000	1.0000	0.7143	0.0000	0.0000
2	1.9600	0.5102	0.4167	0.8167	2.4000	1.2245	0.4167	0.5102
3	2.7440	0.3644	0.2294	0.6294	4.3600	1.5889	0.7798	1.2391
4	3.8416	0.2603	0.1408	0.5408	7.1040	1.8492	1.0923	2.0200
5	5.3782	0.1859	0.0914	0.4914	10.9456	2.0352	1.3580	2.7637
6	7.5295	0.1328	0.0613	0.4613	16.3238	2.1680	1.5811	3.4278
7	10.5414	0.0949	0.0419	0.4419	23.8534	2.2628	1.7664	3.9970
8	14.7579	0.0678	0.0291	0.4291	34.3947	2.3306	1.9185	4.4713
9	20.6610	0.0484	0.0203	0.4203	49.1526	2.3790	2.0422	4.8585
10	28.9255	0.0346	0.0143	0.4143	69.8137	2.4136	2.1419	5.1696
11	40.4957	0.0247	0.0101	0.4101	98.7391	2.4383	2.2215	5.4166
12	56.6939	0.0176	0.0072	0.4072	139.2348	2.4559	2.2845	5.6106
13	79.3715	0.0126	0.0051	0.4051	195.9287	2.4685	2.3341	5.7618
14	111.1201	0.0090	0.0036	0.4036	275.3002	2.4775	2.3729	5.8788
15	155.5681	0.0064	0.0026	0.4026	386.4202	2.4839	2.4030	5.9688
16	217.7953	0.0046	0.0018	0.4018	541.9883	2.4885	2.4262	6.0376
17	304.9135	0.0033	0.0013	0.4013	759.7837	2.4918	2.4441	6.0901
19	597.6304	0.0017	0.0007	0.4007	1491.5760	2.4958	2.4682	6.1601
20	836.6826	0.0012	0.0005	0.4005	2089.2064	2.4970	2.4761	6.1828
21	1171.3556	0.0009	0.0003	0.4003	2925.8889	2.4979	2.4821	6.1998
23	2295.8569	0.0004	0.0002	0.4002	5737.1423	2.4989	2.4900	6.2222
24	3214.1997	0.0003	0.0001	0.4001	8032.9993	2.4992	2.4925	6.2294
25	4499.8796	0.0002	0.0001	0.4001	11247.1990	2.4994	2.4944	6.2347
27	8819.7640	0.0001	0.0000	0.4000	22046.9099	2.4997	2.4969	6.2416
28	12347.6696	0.0001	0.0000	0.4000	30866.6739	2.4998	2.4977	6.2438
29	17286.7374	0.0001	0.0000	0.4000	43214.3435	2.4999	2.4983	6.2454
30	24201.4324	0.0000	0.0000	0.4000	60501.0809	2.4999	2.4988	6.2466
35	130161.1116	0.0000	0.0000	0.4000	325400.2789	2.5000	2.4997	6.2493
36	182225.5562	0.0000	0.0000	0.4000	455561.3904	2.5000	2.4998	6.2495
40	700037.6966	0.0000	0.0000	0.4000	1750091.7415	2.5000	2.4999	6.2498

Appendix B. Solutions

Chapter 1

2. https://studentaid.gov/understand-aid/types/loans/interest-rates

4. https://www.talent.com/tax-calculator?salary=100000&from=year&region=New+York

   https://www.talent.com/tax-calculator?salary=100000&from=year&region=Texas

   The main difference is related to state tax.

6. Answers

   a. This problem is low complexity and may not require financial analysis.

   b. This problem is low complexity and may not require financial analysis.

   c. This problem is intermediate complexity and requires financial analysis.

8. $32,352.94

10.

	Periodicity	Amount variability	Business nature	Opportunity	Production	Tax
Management salary	Fixed	Fixed	Internal	Opportunity	Average	Non-depreciable
Coffee with a discount over volume	Variable	Variable	Internal	Opportunity	Incremental	Non-depreciable
State safety inspection and operation permit	Fixed	Variable	External	Opportunity	Average	Non depreciable
Fine and lawyer cost	Variable	Variable	External	Opportunity	Average	Non-depreciable
Buying a delivery truck every 3 years	Variable	Variable	External	Opportunity	Average	Depreciable
Developed an expansion analysis 2 years ago	Variable	Variable	External	Sunk	Average	Non-depreciable

12. The car manufacturer could have failed with the following:

  a. Misrepresenting marketing information

  b. Avoiding deceptive acts

  c. Failing to report this problem to appropriate professional bodies

Chapter 2

2.

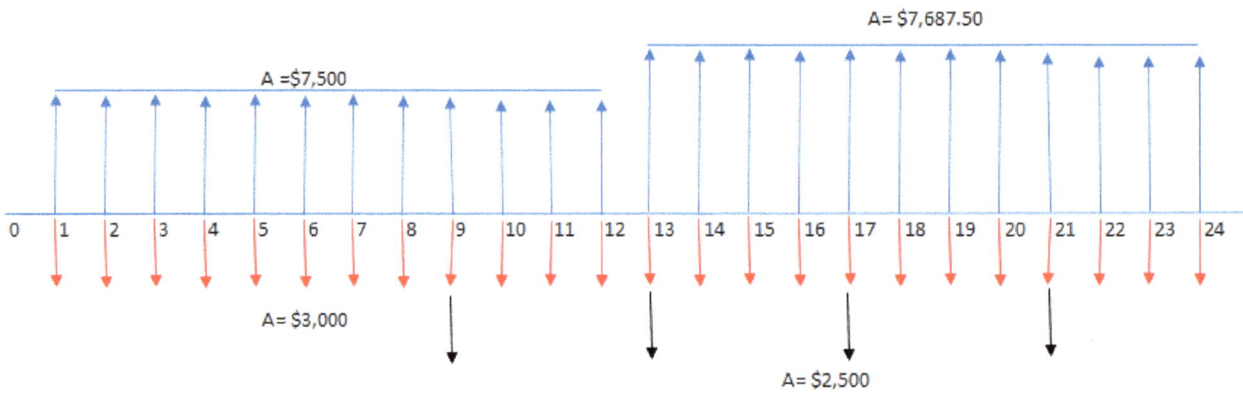

4. $8,640

6.

  a. 25 periods

  b. 100 periods

8. $2,364.38

10. n=23.28 periods

12.

Period (n)	Principal at the start of the period	Interest amount (iA)	End-of-month balance	Interest payment	Principal payment	Total payment
1	$24,000	$720	$24,720	$360	$4,000	$4,360
2	$20,360	$611	$20,971	$305	$4,000	$4,305
3	$16,665	$500	$17,165	$250	$4,000	$4,250
4	$12,915	$387	$13,303	$194	$4,000	$4,194
5	$9,109	$273	$9,382	$137	$4,000	$4,137
6	$5,246	$157	$5,403	$79	$4,000	$4,079
7	$1,324					

14. $74,969.92

16. $223,544.20

18. $100,483.07

20. $2,120,488.09

22. 8.67%

24. 0.9% bi-weekly = effective 26.23%

    2% monthly = effective 26.82%

26. 171.8281%

28. Interest amount =$75

    Rate per period = 257.14%

    Nominal rate = 927.66%

    Future value = $3,596.82

30. Interest amount =$5,000

    Rate per 3-year period = 29.41%

    Rate per 1-year period = 9.80%

    Therefore, borrowing from her uncle (5%) would be cheaper than borrowing from the bank.

Chapter 3

2. $149,959.55

4. −$65,077.20

6. $52,371.22

8. $5,026.07

10. a) $329,569.24

    b) $316,242.89, the amount received by the SSA was greater than what it paid

12. −$522.95

14. a) $286,103.77

    b) $285,611.76

16. a) FV_{15} = $195,209.25 {increase by 7%}

    b) FV_{15} =$198,099.65 {increase by $500}

18. The FV_{25} = $814,114.90

20. X=$25,075.54

Chapter 4

2. $110.40
4. $225,930.35
6. Option a $188,625.13

   Option b $165,419.67
8. $16,268.51
10. $375,000
12. $9,129.12
14. −$15,096.58
16. $32,602.78

Chapter 5

2. a) $2,884.31, no salvage value

   b) $3,103.92, with salvage value
4. EUAC = $650.12
6. a) $8,003,638.41

   b) $10,000,000.00
8. a) $21,556.81

   b) $28,544.01
10. PV= $31,288.75 | EUAC=$4,662.95
12. A PV= $34,576.46 | EUAC $5,152.91
14. a) Payment= $1,827.79

    b) Balance = $24,817.17

    c) Payments = 12.95 instead of 16

Chapter 6

2. i = 6.93%
4. i = 3.11%
6. A = 5.08%

B = 14.27%

Since these alternatives are less than the MARR = 15%, both alternatives are rejected.

8. i per period = 1.92%

   effective i per year = 25.60%

10. i = 7.05%

Chapter 7

2. F= 40.98 A

4. F = $25,232.86

6. Alt A = 14%

   Alt B = 10 %

   Alt C = 18 %

   Alt A- C = -2% Therefore, do Alt C

8. Alt A = 5 years

   Alt B = 6.19 years

   Select Alt A

10. X = 13.61

12. EV = $271.35

14. EV = $5,700

Chapter 8

2. Original Cost $3,500,000

Year	Marcs %	Marcs deprec.
Year 1	0.1429	$500,150
Year 2	0.2449	$857,150
Year 3	0.1749	$612,150
Year 4	0.1249	$437,150
Year 5	0.0893	$312,550
Year 6	0.0892	$312,200

Year 7	0.0893	$312,550
Year 8	0.0446	$156,100

4. a) depreciation recapture $50,000

   b) depreciation recapture $15,000

   c) loss $20,000

6. Texas has 0% corporate tax. The combined tax rate is 21%

   Illinois has a 9.5% corporate tax. The combined tax rate is 28.51%

   Florida has a 5.5% corporate tax. The combined tax rate is 25.35%

   New York has a 7.25% corporate tax. The combined tax rate is 26.93%

8. a) depreciation for year 1, $400,000 for year 2, $640,000

   b) Taxable income year 1 $3,100,000 year 2 $8,560,000

   c) Combined tax rate 26.93%

10. A

12. A

14. A

16. A

18. A

Credits

IMG 1.1: Copyright © 2018 Depositphotos/svyatoslavlipik.

IMG 2.1: Copyright © 2013 Depositphotos/xxxPATRIK.

Fig. 2.1: Copyright © 2011 Depositphotos/georgios.

Fig. 2.7: Copyright © by Microsoft.

Fig. 2.8: Copyright © by Microsoft.

Fig. 2.9: Copyright © by Microsoft.

Fig. 2.10: Copyright © by Microsoft.

IMG 3.1: Copyright © 2015 Depositphotos/Daxiao_Productions.

Fig. 3.4: Copyright © by Microsoft.

Fig. 3.5: Copyright © by Microsoft.

Fig. 3.6: Copyright © by Microsoft.

Fig. 3.7: Copyright © by Microsoft.

Fig. 3.8: Copyright © by Microsoft.

Fig. 3.10: Copyright © by Microsoft.

Fig. 3.12: Copyright © by Microsoft.

Fig. 3.14: Copyright © by Microsoft.

IMG 4.1: Copyright © 2011 Depositphotos/iriana88w.

Fig. 4.10: Copyright © by Microsoft.

Fig. 4.11: Copyright © by Microsoft.

Fig. 4.12: Copyright © by Microsoft.

Fig. 4.13: Copyright © by Microsoft.

Fig. 4.14: Copyright © by Microsoft.

Fig. 4.15: Copyright © by Microsoft.

Fig. 4.16: Copyright © by Microsoft.

IMG 5.1: Copyright © 2011 Depositphotos/lightpoet.

Fig. 5.6: Copyright © by Microsoft.

Fig. 5.7: Copyright © by Microsoft.

Fig. 5.8: Copyright © by Microsoft.

Fig. 5.9: Copyright © by Microsoft.

Fig. 5.10: Copyright © by Microsoft.

Fig. 5.11: Copyright © by Microsoft.

Fig. 5.12: Copyright © by Microsoft.

Fig. 5.13: Copyright © by Microsoft.

IMG 6.1: Copyright © 2013 Depositphotos/kzlobastov.

IMG 6.2: Copyright © 2017 Depositphotos/odua.

Fig. 6.1: Copyright © by Microsoft.

Fig. 6.2: Copyright © by Microsoft.

Fig. 6.3: Copyright © by Microsoft.

Fig. 6.6: Copyright © by Microsoft.

Fig. 6.7: Copyright © by Microsoft.

IMG 6.3: Copyright © by Microsoft.

Fig. 6.8: Copyright © by Microsoft.

Fig. 6.9: Copyright © by Microsoft.

Fig. 6.10: Copyright © by Microsoft.

Fig. 6.11: Copyright © by Microsoft.

Fig. 6.12: Copyright © by Microsoft.

Fig. 6.13: Copyright © by Microsoft.

Fig. 6.14: Copyright © by Microsoft.

Fig. 6.15: Copyright © by Microsoft.

Fig. 6.16: Copyright © by Microsoft.

Fig. 6.17: Copyright © by Microsoft.

Fig. 6.18: Copyright © by Microsoft.

Fig. 6.19: Copyright © by Microsoft.

Fig. 6.20: Copyright © by Microsoft.

Fig. 6.21: Copyright © by Microsoft.

IMG 7.1: Copyright © 2013 Depositphotos/creatista.

IMG 7.2: Copyright © by ABC Television Network.

IMG 7.3: Copyright © 2011 Depositphotos/toxawww.

IMG 7.4: Copyright © 2020 Depositphotos/BiancoBlue.

IMG 7.5: Copyright © 2018 Depositphotos/denisismagilov.

Fig. 7.5a: Copyright © 2021 Depositphotos/olgacov.

IMG 7.7: Copyright © 2015 Depositphotos/orlaimagen.

Fig. 7.8: Copyright © 2020 Depositphotos/Furian.

IMG 8.1: Copyright © 2013 Depositphotos/stokkete.

IMG 8.2: Copyright © 2019 Depositphotos/Sonsedskaya.

www.ingramcontent.com/pod-product-compliance
Ingram Content Group UK Ltd.
Pitfield, Milton Keynes, MK11 3LW, UK
UKHW052011041026
15590UKWH00008B/2269

9 781793 524454